Questionnaire of Sugarcane & Quality Control

Questionnaire of Sugarcane & Quality Control

Dr. B.S. Tomer
Vijay Singh

Notion Press

Old No. 38, New No. 6
McNichols Road, Chetpet
Chennai - 600 031

First Published by Notion Press 2016
Copyright © Dr. B.S. Tomer, Vijay Singh 2016
All Rights Reserved.

ISBN 978-1-945579-27-1

This book has been published with all efforts taken to make the material error-free after the consent of the author. However, the author and the publisher do not assume and hereby disclaim any liability to any party for any loss, damage, or disruption caused by errors or omissions, whether such errors or omissions result from negligence, accident, or any other cause.

No part of this book may be used, reproduced in any manner whatsoever without written permission from the author, except in the case of brief quotations embodied in critical articles and reviews.

DEDICATED

TO

OUR PARENTS

CONTENTS

Preface	*xiii*
Abbreviations	*xv*
Part I: General Questions on Sugarcane	**3**
1. Sugarcane Classification	9
2. Land Preparation	10
3. Seed Selection and Treatment	11
4. Planting Techniques	17
5. Sugarcane Varieties	23
6. Nutrient Management	25
7. Soil and Soil Health	35
8. Weed Management	41
9. Foliar Spray on Sugarcane	46
10. Diseases of Sugarcane	51
11. Insect Pest Management	61
12. Bio Fertilizer	79
13. Intercropping and Crop Rotation	82
14. Ratoon Management	85
15. Drip Irrigation in Sugarcane	88
16. Advantage of Neem in Sugarcane	91
17. Cane Marketing in Uttar Pradesh	93
18. Agronomical Practices in Kenya	98
Part II: Basic Quality Control	**103**
1. Quality Management System	127
2. Food Safety Management System	170
3. Sample Paper for Officer (Cane)	199
References	*207*

डॉ० संजीव कुमार बालियान
DR. SANJEEV KUMAR BALYAN

कृषि एवं किसान कल्याण राज्य मंत्री
भारत सरकार
Minister of State for Agriculture
and Farmers Welfare
Government of India

FOREWORD

Sugarcane is a one of the major crop in India as well as some other countries like Australia, Thailand, Kenya, Uganda, South Africa, Brazil, Philippines etc. It has big share in Indian economy because 528 sugar factory are in working condition. If cane productivity per unit area increases then grower would be happy with better returns. The latest technology in agriculture which has been developed by agricultural scientists, if reach "Lab to Land" then the productivity is bound to increase. Authors has attempted to produce all scientific technologies in a hand book question-answer pattern in easy language namely "QUESTIONNAIRE OF SUGARCANE & QUALITY CONTROL".

It is hoped that this book will be good medium to produce the latest information to sugar industry professional and agricultural students as well as growers.

With best wishes.

Date: 20.04.2016

(Dr. Sanjeev Balyan)

199-Q, Krishi Bhawan, New Delhi-110 001 Tel.: 91-11-23782343 Fax: 91-11-23074190

SAMEER SINHA
PRESIDENT

ENGINEERING & INDUSTRIES LTD.

CORPORATE OFFICE
8th Floor, Express Trade Towers, 15-16, Sector 16A, Noida - 201301, U.P. India
T: +91 120 4308000 | F: +91 120 4311010-11
W: www.trivenigroup.com

FOREWORD

Sugar Industry is the second largest agro processing industry in India and has approximately 40 million farmer families dependent on it. The leading sugarcane producing states in India are Maharashtra, UP, Karnataka, Tamil Nadu and Andhra Pradesh etc. While India is yet to catch up with the best internationally in terms of productivity & recovery, even within the country, there are wide variances in these parameters across various regions, indicating potential for improvement all across and specially in many of the underperforming regions.

Therefore, the latest developments & techniques which have been developed by the agricultural scientists must reach the grower and be practised by him. This will result in better returns to all the stakeholders viz. farmers, millers etc.

The authors in this book "QUESTIONNAIRE OF SUGARCANE & QUALITY CONTROL." have attempted to compile the technologies, developments, techniques related to sugarcane in a lucid question – answer format for easy comprehensibility & understanding and thereby assist in ensuring that such developments & techniques are taken from "lab to land" resulting in such practices being followed by the growers on a sustainable basis for superior performance.

I am happy to note that the first author has published 25 research papers on different crops including 5 papers in International journals besides a previous book on sugarcane titled "Ganna Prashanottary" in Hindi. The authors have excellent cumulative experience in the sugar sector having worked with the leading and most reputed sugar manufacturing organizations in India which have carried out pioneering work in this field.

I congratulate the authors for their work and I am confident that this book will be extremely useful for sugar industry professionals, agricultural students, sugarcane growers and will assist in bringing agricultural students more up-to-date in terms of knowledge as well as make the grower more aware and knowledgeable about the best practices and thus bridge the gap between the growers and the sugarcane development professionals of the sugar mills.

Date: 11.04.2016

(SAMEER SINHA)

Regd. Office: Deoband, District Saharanpur, Uttar Pradesh - 247 554
CIN No. L15421UP1932PLC022174

West Kenya Sugar Company Limited

P. O. Box 2101, Kakamega, Kenya
Telephone : +254 (020) 2036320/30/40   Fax : +254 (020) 2036370/80
Mobile   : +254 (0722) 786084/786404 Email: info@wksugar.com
Factory : South Kabras, Kakamega North District

KENYA'S SWEETEST
ISO 9001 - 2008 Certified

KARAN SINGH
GENERAL MANAGER

<u>FOREWORD</u>

Sugarcane is a one of the major crop in India as well as some other countries like Australia, Thailand, Kenya, Uganda, South Africa, Brazil, Philippines etc. It has big share in Indian economy because 528 sugar factory are in working condition whereas 126 are only in Uttar Pradesh. The sugarcane productivity is less in U.P. as compared to Tamilnadu, Karnataka and Maharashtra, however the cane area is more in the state. If cane productivity per unit area increases then grower would be happy with better returns. The latest technology in agriculture which has been developed by agricultural scientists, if reach "Lab to Land" then the productivity is bound to increase. Authors has attempted to produce all scientific technologies in a hand book question-answer pattern in easy language namely "QUESTIONNAIRE OF SUGARCANE & QUALITY CONTROL".

The author has also published one hand book in 2005 in Hindi language namely "Ganna Prashanottary." That book was very useful to all field staffs/officers of sugar factories as well as sugarcane growers, because that has proved to very informative for updating the knowledge.

It is hoped that this book will serve as repository of information and will be highly useful for sugar persons, agricultural students and progressive growers. May this book fulfill all gaps between sugar mill cane professionals and growers.

With best wishes.

Date: 16.03.2016

(Karan Singh)

DIRECTORS : Jaswant S. Rai (Chairman) | Sarbjit S. Rai (Director) | Tejveer S. Rai (Managing)

PREFACE

This book offers guidelines on sugarcane development and quality control, quality management systems and food safety management in the sugar industry.

We are thankful to everyone who provided support material for this manuscript: the scientists of Sugarcane Breeding Institute (SBI), Uttar Pradesh Council of Sugarcane Research (UPCSR), Indian Institute of Sugarcane Research (IISR) and National Sugar Institute (NSI) and the scientists involved in developing new sugarcane varieties, new technologies with new agricultural implements and organic liquid fertilizers for better quality and yield and quality system improvement.

Dr. Tomer is heartily thankful to all those people responsible for the success of his first book *Ganna Prashanottary*. Among the several people who helped in the compilation of this book, particular mention has to be made of Late Shri Mangal Singh, who was the motivator (director) of Dr. Tomer's first book. Another person whose invaluable contribution Dr. Tomer would like to acknowledge is his educational boss, (Professor) Dr. S C Goel.

We trust that this book, besides its intended purpose, will also help in enhancing the knowledge base of practicing cane professionals and sugar technologists as well as agricultural students.

We acknowledge that this book has scope for further improvement and invite your valuable suggestions and comments, which may be mailed to drbstomer@gmail.com and vij2singh@gmail.com

Last but not the least, the authors would like to thank their family members who always support and encourage new endeavors.

DR. B S TOMER   VIJAY SINGH

ABBREVIATIONS

Al	Aluminum
ANSI	American National Standards Institute
ASQC	American Society for Quality Control
BH	Boiling House
Bo	Boron
BRC	British Retailer Consortium
BSI	British Standards Institution
BIS	Bureau of Indian Standard
Ca	Calcium
CASCO	ISO Committee on Conformity Assessment
CCP	Critical Control Point
CIC	Cane Implementing Committee
CIP	Cleaning in Process
Cl	Chlorine
COP	Cleaning out of place
Cu	Copper
CV	Coefficient of Variation
DaP	Days After Planting
DAP	Diammonium Phosphate
DCO	District Cane Officer
DIS	Draft International Standard
EIL	Economic Injury Level
ESB	Early Shoot Borer
ET	Economic Threshold
EU	European Union
FDF	Food and Drink Federation
Fe	Iron
FEFO	First Expired First Out
FIFO	First In First Out

FIRB	Furrow-irrigated raised-bed
FMEA	Failure Mode Effects Analysis
FSMS	Food Safety Management System
FSSC	Food Safety System Certification
FSTL	Food Safety Team Leader
FYM	Farm Yard Manure
GA	Gibberelic Acid
GAP	Good Agricultural Practices
GDP	Good Documentation Practices
GFSI	Global Food Safety Initiatives
GHP	Good Hygiene Practices
GLP	Good Laboratory Practices
GMP	Good Manufacturing Practices
GS	General Specified
GSD	Grassy Shoot Disease
GSP	Good Sanitary Practices
HA	Hazard Analysis
HACCP	Hazard Analysis Critical Control Point
HHC	Hand-Held Computer
HLS	High-Level Structure
HoD	Head of Department
HWT	Hot Water Treatment
IA	Internal Audit
IAA	Indol Acitic Acid
IAF	International Accreditation Forum
ICUMSA	International Commission for Uniform Methods of Sugar Analysis
IFS	International Food System
IISR	Indian Institute of Sugarcane Research
IPM	Integrated Pest Management
ISO	International Organization of Standardization
ISO/TC 176	ISO Technical Committee No 176
IU	ICUMSA Units

Abbreviations

KALRO-SRI	Kenya Agricultural & Livestock Research Organization- Sugarcane Research Institute
MA	Mean Apparatus
MB	Moldboard
Mg	Magnesium
ML	Milliliters
MM	Millimeters
Mn	Manganese
Mo	Molybdenum
MR	Management Representative
MRM	Management Review Meeting
MSDS	Material Safety Data Sheet
NPK	Nitrogen, Phosphorus, Potassium
NSI	National Sugar Institute
OLF	Organic Liquid Fertilizer
OPRP	Operational Prerequisite Program
P	Phosphorus
PAS	Publicly Available Specification
PDCA	Plan-Do-Check-Act
PI	Preparatory Index
PRP	Pre-Requisite Program
QA	Quality Assurance
QC	Quality Control
QI	Quality Improvement
QM	Quality Manual
QMS	Quality Management System
QP	Quality Planning
RBS	Raised Bed Seedlings
RMD	Ratoon Management Device
S	Sulphur
SBI	Sugarcane Breeding Institute
SOP	Standard Operating Procedure
SPS	Sanitary and Phytosanitary

Abbreviations

SSI	Sustainable Sugarcane Intensification
SSOP	Sanitation Standard Operating Procedure
STP	Spaced Transplanting
TC	ISO Technical Committee
TQM	Total Quality Management
TS	Technical Standard
UPCSR	Uttar Pradesh Council of Sugarcane Research
WHO	World Health Organization
WI	Work Instructions
Zn	Zinc

PART I

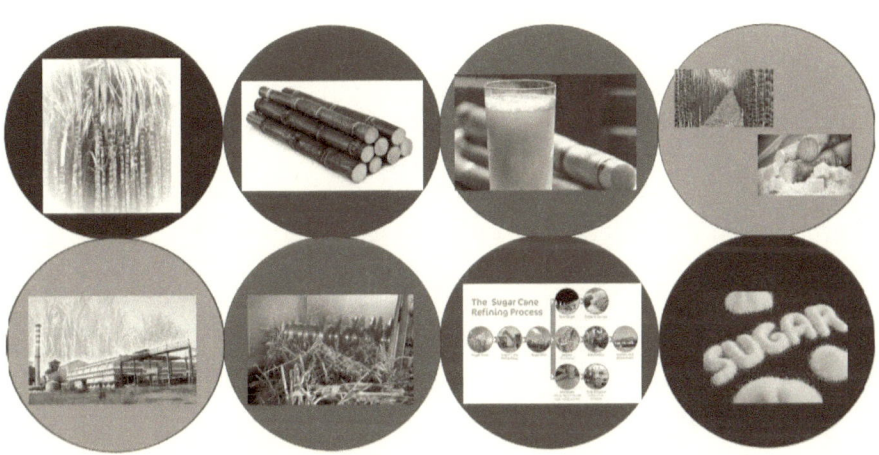

GENERAL QUESTIONS ON SUGARCANE

This chapter contains only those questions that are most common in the industry and must be known by each sugar industry professional and readers must go through the first chapter since it stimulates interest in the subject. They will find, on completing the first chapter that they cannot put down the book without reading it entirely.

Q. What is sugarcane?

ANS. Sugarcane (*Saccharum officinerum*) is a perennial grass that belongs to the "Graminee" family that originated in New Guinea (Asia). "Saccharum" is derived from the Sanskrit "Sakara" meaning white sugar. Sugarcane is a sub-tropical and tropical crop that requires a lot of water and sunlight to store energy. The crop's maturity period is typically about twelve months, but it varies from six to eight months in Louisiana to eighteen–twenty months in Kenya[3] due to rain fed area.

Q. What is the best climate for the optimum growth of sugarcane?

ANS. The optimum temperature range for sugarcane germination is 15–25 °C. It requires high temperatures for growth, with optimum temperatures of 20 to 30 °C. Temperatures below 20 °C affect the length of the growing season. However, low temperatures are the most effective way to ripen cane. Although fluctuations in temperature may have a positive effect on sucrose accumulation, a temperature of less than 5 °C is potentially damaging to growth.[2]

Q. How many stages does a sugarcane crop cycle have?

ANS. Sugarcane attains maturity in five stages: 1. Germination, 2. Tillering, 3. Formative, 4. Grand Growth and 5. Maturity.

Q. What is each crop stage?

ANS. Germination Phase: This is the single most critical factor in determining sugarcane yield. Treating sugarcane setts with 1% solution of Rexil/Bavistin aids the germination process. Dipping the set in urea solution also enhances germination. The seed cane must be watered for seven days before harvesting it from the field of seed. Reducing the time lag between harvesting and planting of seed cane also improves germination. Lime solution is the best option for increasing germination in acidic soil.

Tillering Phase: This stage provides the appropriate number of cane stalks required for good yield. For maximum tillering, the temperature should be around 30 °C and not below 20 °C.

Formative Phase: This requires a temperature of around 30 °C, 50% atmospheric humidity and an abundance of soil moisture. This phase decides the number of millable cane and contributes 40% to cane yield.

Grand Growth Phase: This phase starts from July and ends around mid-October. Sugarcane grows 5 inches per week during this phase. Favorable temperature is 30–35 °C with 80% humidity. As much as 70 to 80% of the weight of the cane is decided in this phase. The length of the cane contributes 30% to cane yield. In the absence of rain, irrigation is required every fifteen-twenty days during this phase.

Maturity Phase: This relates to sugar accumulation in the cane. Vegetative growth is reduced in this phase. Cane ripening starts from the bottom and moves to the top; the bottom portion thus contains more sucrose than the top. A minimum temperature of 7–15°C and a maximum temperature of 20–25°C, that is cool nights and bright days, are ideal for sugar synthesis.[2]

Q. What type of plant is sugarcane?

ANS. Sugarcane, *Saccharum officinerum*, belongs to the grass family, which has fast growth with a C4 carbon cycle and a high chromosome number.

Q. How many byproducts does the sugarcane crop have?

ANS. Sugarcane is a multipurpose crop. It provides animal fodder; sugar press mud can be used as a bio fertilizer; sugar fermentation produces ethanol; sugar is used in antibiotics and in particle board and bagasse can be used for electricity generation.

Q. Why does the yield per unit area have to be increased?

ANS. Cultivable area available for sugarcane is limited, while conversion of agricultural land for industrial and construction purposes continues unabated. Thus, attempts must be made to increase the per unit area yield of sugarcane.

Q. How many chromosomes does sugarcane have?

ANS. Sugarcane does not have a fixed chromosome number like other crops; it is a range, for example, $2n = 40$ to 120.

Q. In how many countries is sugarcane grown?

ANS. Sugarcane is highly adaptable to a wide range of tropical and subtropical climates and soils. Thus, it is grown in over 100 countries.

Q. How many harvest cycles does a crop yield?

ANS. The sugarcane production cycle typically has two to six harvests in most countries; however, it can be extended to over thirty harvests in Swaziland.

Q. What is plant quarantine?

ANS. The interchange of varieties between sugarcane-producing countries is a common practice. It is necessary for the introduction of possibly superior hybrids for commercial production and the exchange of germplasm between institutes that conduct sugarcane-breeding programs. Within the quarantine facility, the imported material is usually grown for two plantings, including a ratoon crop stage, according to protocols that usually include hot water treatment (HWT) to eliminate any bacterial or fungal pathogens.

Q. What is the difference between irrigated and rain-fed sugarcane crops?

ANS. Irrigated crops may be harvested within twelve months, while rain-fed crops can only be harvested after sixteen months of planting.

Q. Why is sugarcane productivity declining?

ANS. Sugarcane productivity is declining due to excess use of water and imbalance in the dosage of fertilizers and improper timeframes.

Q. What are the factors affecting yield?
ANS. Yield is affected by:

1. Unavailability of or carelessness in selection of good quality seed.
2. Lack of proper treatment and cutting of the seed.
3. Unbalanced dose of fertilizers.
4. Lack of knowledge of new technology.
5. Non-consideration of timeframes in applying plant-protection measures.
6. Lack of awareness about best packages and practices.
7. Negligence in ratoon management.
8. Excess use of water and nitrogenous fertilizers.

Q. Which Indian states have the lowest and highest sugarcane productivity?
ANS. Madhya Pradesh has the lowest sugarcane productivity in India at 40 ton/ha while Tamil Nadu has the highest at 135 ton/ha.

Q. What is the average yield of sugarcane-growing countries?
ANS. The global average yield of sugarcane is 59.50 ton/ha.

Table 1: Sugarcane yield per hectare of some countries.

S.No.	Country	Cane yield (MT/ha)	
		2012	2013
1	India	70.93	67.43
2	Brazil	74.29	75.34
3	Thailand	76.75	75.74
4	Indonesia	70.00	74.89
5	Australia	76.65	82.40
6	Uganda	70.21	67.00
7	South Africa	53.99	55.39
8	Pakistan	55.82	56.48
9	Kenya	68.57	69.41

Source: FAO-STAT 2014

Q. Which three countries have the highest area under sugarcane cultivation?

ANS. In 2013, Brazil was first with its area under sugarcane cultivation at 37.84% of the global total, India was second with 18.78% and China was third with 6.77%.

Q. Which three countries lead in sugarcane production?

ANS. In 2013, Brazil was first, with cane production accounting for 40.18% of global cane production, India was second at 17.85% and China was third at 6.74%.

Q. How much of the world's sugar requirement is fulfilled by sugarcane?

ANS. Sugarcane meets about 70% of the world's sugar requirement.[19]

Q. What is the per capita per year demand for sugar?

ANS. Per capita consumption varies from country to country; the global average is 25 kg per person per year.[19]

Q. Who are the main producers of sugar?

ANS. The three major producers are Brazil, which produced 33.78 million MT, India which produced 28.20 million MT, and the European Union (EU), which produced 18.06 million MT, in 2014–15.[19]

Q. Which countries are the top three raw sugar exporters?

ANS. Brazil, Australia and Thailand.

Q. Which countries are the top three raw sugar importers?

ANS. Russia, the European Union and USA.

Q. How much water is consumed to produce one kilogram of sugar?

ANS. To produce one kilogram of sugar, 4.5–5.0 liters of water are required.

Q. What are the main byproducts of the sugar industry?

ANS. The main byproducts of the sugar industry are molasses, bagasse and filter cake.

Q. What is molasses?

ANS. Molasses is the main raw material for alcohol-based products (ethanol, chemicals) and it is also used in animal feed.

Q. What is bagasse?

ANS. Bagasse is the main raw material for the paper industry. It is also used as a fuel in the generation of power.

Q. What is filter cake?

ANS. Filter cake is mainly used as manure in fields but over the past few decades, it has also been used in the brick industry along with coal.

Q. Is sugarcane helpful in ethanol production?

ANS. In the current decade, sugarcane has emerged as the most efficient source of ethanol. Globally, the average production of ethanol is 80–85 liters per ton of sugarcane. Ethanol production depends mainly on the sugar content in the cane.

Q. Which country leads in ethanol production?

ANS. Brazil has the highest level of ethanol production from sugarcane, that is, 200 million liters from 30,000 hectares of sugarcane.

Q. How do we measure rain?

ANS. An instrument called a rain gauze is used to measure rain. It records the amount of rain that has fallen during a particular length of time. A rain gauze usually measures rainfall in millimeters (mm), although some rain gauze cylinders measure in milliliters (ml) and then convert it into millimeters as under:

$$ml \times 0.254 = \underline{} mm.$$

Q. How can a rain gauze be built at home?

ANS. A rain gauze can be built at home using a one-liter plastic bottle, a graduated cylinder, a funnel and a permanent marker. Cut the top of the plastic bottle. Using the graduated cylinder, pour 10 ml of water into the bottle. Each 10 ml can be marked with a permanent marker, with sub columns up to the end of bottle. Then, empty the bottle, place the funnel on top of it and keep it outside to measure rainfall.

SUGARCANE CLASSIFICATION

This chapter provides information about the sugarcane that would be considered basic knowledge for any cane professional.

Q. How many species of sugarcane have been identified?

ANS. Six species of sugarcane have been identified. Of these, only three are capable of producing cane or being used for cropping. The rest are wild species of sugarcane. The details of the different species are as under:

1. *Saccharum officinerum*: The cane of this species is known as Noble cane. It originated in New Guinea and is bright in color, thick and soft. It has a higher percentage of sucrose.

2. *Saccharum sinense*: The cane of this species is known as China cane, because it originates from China. This cane has a high fiber content, but it is thinner, harder and has lesser sucrose than Noble cane.

3. *Saccharum barberi*: The cane of this species is known as Indian cane. It has moderate sucrose content but can better withstand drought and frost (pala).

4. *Saccharum spontaneum*: This species of cane is thin, long like a pipe and green-yellow in color. As it ripens, it turns yellow or white in color.

5. *Saccharum robustum*: This is a purely wild species with a cane of medium thickness and very low sucrose content.

6. *Saccharum edule*: This shares a lot of similarities with *robustum* and is known as a wild species.

LAND PREPARATION

The information provided in this chapter can be applied by sugarcane growers to get a good yield. Sometimes, growers do not care about fine tilth, without which the root zone cannot develop as per the plant's requirement.

Q. What is the importance of land preparation to sugarcane yield?

ANS. Sugarcane requires well-prepared soil to ensure sufficient moisture retention. The soil must have good tilth to facilitate good furrow and earthing up.

Q. How can this type of soil be prepared?

ANS. To prepare soil, deep ploughing with a disc or Mold Board (MB) plough is required. The soil must be harrowed a minimum of two times with two cross tillering if possible. It must be ploughed with a rotavator for good tilth. Thereafter, planking must be undertaken.

First, use an MB Plough or chiseler; second, use a disc plough (for cross tillage); third, use a harrow (for cross tillage) and fourth, use a rotavator.

Furrowing must be with a minimum distance of 90 cm; for a high yield, it can be 120 cm apart but seed rate should not be less than 3.5–4.0 ton per acre.

Q. Why is good tilth required for a sugarcane crop?

ANS. Since sugarcane is a perennial crop, once planted it can grow in the field for three or four years. Unwanted vegetation grows in the field very fast, so to control weeds and help the sugarcane root system develop well, good tilth is required.

SEED SELECTION AND TREATMENT

In the current scenario, since agricultural labor is not easily available, growers plant their crop without seed treatment. Due to this, the life of cane varieties becomes short owing to unwanted fungal infestation and the variety degrades before time.

Q. Why is good quality seed required?

ANS. Good quality seed is the base of any crop. As the Hindi proverb says, "*Jaisa boyenge waisa katenge*," you reap what you sow.

Q. Can sugarcane seed be taken from a regular field?

ANS. No, never, because regular (commercial) crops have all kinds of cane including dried cane, other varieties of cane and diseased cane.

Q. How can germination be increased?

ANS. When planting rice, a nursery can be planted from a selected seed, but with sugarcane— where there is a shortage of time, labor, money, etc.—for a good and healthy crop, the seed must always be selected from a nursery that has been properly checked by cane officials. After selecting and harvesting the seed, it must be treated with a 1% fungicide (Rexil or Bavistin) solution to prevent fungal disease and meet the water requirement of the seed. This will ensure good germination.

Fig. 1

Q. What measures must be taken if sugarcane is to be planted in rain-fed areas or areas with minimum irrigation?

ANS. In rain-fed areas or where irrigation is limited, seed cane must be treated with lime solution and 90 DaP. Nitrogen, Phosphorus and Potash (NPK)/Potassium chloride foliar spray can be used but the field should be watered before the spray is used.

Q. How can the age of seed cane be verified?

ANS. Seed cane of only nine-ten months of age should be selected for good germination. If this is not possible, then only the upper one-third portion of a cane should be planted because this has less sucrose. The age of the seed cane can be verified by counting internodes. For example, select a cane from a bundle or field and count its internodes.

Suppose the internodes are 15, then

Age of cane = $15 \times 2/3$

= 10 months

Q. What are the parameters for checking a seed cane nursery?

ANS. To select seed cane, the following parameters must be inspected by concerned officials:

a. disease, b. insect pests, c. growth parameters, d. uniformity and e. damage caused by rodents/wild animals.

A report must then be submitted in the following format:

1. Name of Farmer and Father's Name; 2. Cane Variety; 3. Date of Planting; 4. Contact No; 5. Fertilizers, Quantity and Time; 6. Expected Yield (MT/ha); 7. Name of Field Assistant; 8. Zone; 9. Village; 10. Circle. Remarks: Recommendations for seed, suitable/not suitable. Signature of checking officer.

Q. Why must urea be applied in the seed nursery before harvesting?

ANS. For each acre in the nursery, 25–30 kg of urea can be applied one week before harvesting the seed. Urea breaks the sucrose into three molecules of glucose and fructose. The eye bud germinates only in the presence of glucose in seed setts. If urea is applied, the percentage of germination increases in lesser time.

Q. What is the alternative if urea cannot be used in the nursery?

ANS. If urea cannot be applied to the standing crop, then, if possible, harvest the seed, cut it into setts, then dump it in a canal filled with water. To this water-filled canal, add urea and fungicide and leave for at least 10–15 minutes. Alternatively, treat the seed setts with a urea and fungicide mixture @ 1% solution in a drum.

Q. Why must the seed cane be treated with fungicide?

ANS. When the seed cane is treated with fungicide, then we secure it against infestation from any fungus. On the seed cane, both the ends are cut and open so the fungus spore can enter and attack.

Q. What impact does gibberellic acid in fungicide solution have on germination?

ANS. When seed cane is treated by fungicide with gibberellic acid @ 1 gm/acre, maximum germination can be achieved with greater tillering capacity. However, for this treatment, the distance between rows must be 100 to 120 cm because after treatment, the crop attains more tillers.

Q. If a field is prone to termite and white grub infestation, how can it be secured?

ANS. If a field is termite infected, then the seed cane must be treated with imidacloprid (Confidor) @ 125 ml/acre in 125 liters of water, or Lesenta @ 150 gm/acre dissolved in 400 liters of water. The solution must be drenched on the seed cane in lines.

Q. What is the bio control method that can be used against white grub and termites?

ANS. *Buevaria bassiana* @ 2 kg/acre and *metaryzium anasopli* @ 2 kg/acre prepared in a culture of decomposed cow dung must be left for thirty–forty days. Thereafter, the culture would be ready for use against white grub and termite. At the time of land preparation, this culture can be used in a wet field. During planting, precaution must be taken to ensure that the chemical has not been left for more than forty days. This culture can be repeated in July after weeding/inter culturing.

Q. What is a home remedy to control termites?

ANS. In a termite-infested field, dig a pit of 1 × 1 × 1 feet, each at a distance of 10–15 meters from the boundary and fill it with raw cow dung for at least twenty–thirty days. Alternatively, take an earthenware pot (*ghada*) and make a small hole in it. Fill it with fresh cow dung. Dig in each corner and in the center of the field and leave the pot under the soil for twenty–thirty days. After this period, burn the cow dung with kerosene or wash it away with water (pond) because most of the termites of the field would be in the cow dung. This method can be repeated two or three times in an infested field as the chief and best method for controlling termites.

Q. How can neem help in termite control?

ANS. Five kg of green neem leaves and 2 kg of *madaar (aakh)* in an earthenware pot (*ghada*) with a capacity of 25–30 liters filled with 15 liters of cow urine must be kept tightly shut for fifteen days. Thereafter, the organic insecticide will be ready and can be used on any crop against any insect for foliar or soil application. This 25–30 liter mixture may be diluted in 100 liters of water and drenched in furrows on the seed to protect against termite and white grub.

Q. What is a nursery?

ANS. A nursery is a first step for any crop. It is used for propagation of seed. Different types of sugarcane nurseries are as under:

1. *Adhar* nursery (foundation),
2. Primary nursery, and
3. Secondary nursery.

Q. What is an *adhar* nursery?

ANS. After releasing the new cane variety, the seed quantity that is obtained from the research station for the nursery is called *adhar* nursery (foundation nursery).

Q. What is a primary nursery?

ANS. The nursery established from the seed of the *adhar* nursery is called a primary nursery.

Q. How can rapid multiplication be achieved with fewer seed of new cane variety?

ANS. When only limited quantity of seed is available, packet nursery technique must be applied at the progressive farmer's field in a factory reserved area.

Q. What is packet nursery technique?

ANS. Packet nursery is a very effective technique for rapid multiplication of new seed varieties in the factory command area. In this technique, treated (fungicide and insecticide) seed setts @ 25–50 kg are provided to a progressive farmer with the cost deducted from cane price payment. Before providing the seed, the grower must be educated about it. Factory officials must visit all nurseries fortnightly and thoroughly check the soil quality and look for pest/disease infestation. If this is noticed, then timely control measures must be adopted.

Q. How can yield be increased with agronomical activities?

ANS. To increase yield with agronomical activities, the following steps may be taken:

1. Land must be prepared with fine tilth for root development.
2. Row-to-row space should not be less than 90 cm; for greater single cane weight, this must be increased to 120–150 cm.
3. Farm Yard Manure (FYM) must be applied @ 10 ton/acre at the time of land preparation and a basal dose of inorganic fertilizers is needed before seed is set in furrow, that is, below the seed in recommended dose.
4. Seed must be disease- and pest-free and only two/three budded sets must be used.
5. During planting, the seed must be treated with a 1% solution of fungicide and urea to increase germination and protect from fungal infection.
6. Termite/white grub prone areas must be treated with insecticide such as Lesenta/Dontatsu @ 150 gm/acre in 400 liters of water. This must be used to drench the seed in furrow.

7. First weeding must be done one–three days after planting (DaP) with weedicide, second weeding with power rotavator must be after completion of germination and then the first dose of top dressing nitrogenous fertilizers and earthing up can be conducted. Third weeding must be 100–120 DaP with power tiller along with second dose of fertilizer and earthing up.

8. Mechanical weeding not only removes weeds but also increases sprouting tillers and enhances aeration in the soil.

PLANTING TECHNIQUES

Cane professionals can educate growers about various planting techniques and their benefits. Growers, in turn, can apply these techniques in their fields for greater yield at lesser cost.

Q. What factors affect the yield of sugarcane?

ANS. The factors affecting sugarcane yield are as under:

1. New technologies,
2. Seed quality,
3. Soil,
4. Climate, and
5. Human resources

Q. What are the different sugarcane-planting techniques?

ANS. Details of different sugarcane-planting techniques are as under:

1. **Furrow method planting:** This method is commonly adopted by growers, especially in low soil moisture conditions and involves planting mostly three-budded sets of seed kept in deep (10–15 cm) furrows 90 cm apart. The furrows are covered with 5–6 cm of soil, and irrigation is conducted in the furrow.

2. **Mechanized furrow planting:** With the three-row planter developed by IISR, Lucknow, just five persons can plant one acre in two–three hours.

Fig. 2

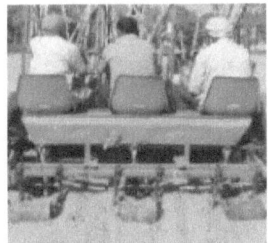

3. **Paired row method planting**: Paired rows (two furrows) are 60 cm apart, with 90 to 120 cm distance between two pairs. Two or three budded sets are kept 10–15 cm deep in the furrows. This method increases the amount of millable cane and keeps it healthier than in furrow method planting. This is most profitable for the intercropping pattern of sugarcane planting.

4. **FIRB planting**: Furrow-Irrigated Raised-Bed (FIRB) planting is most beneficial in wheat–sugarcane, sugarcane–peas, sugarcane–lentil, sugarcane–gram, etc. In this method, tractor operated machines drill the seed of wheat and fertilizers into the bed and open the furrow for sugarcane.

Fig. 3

5. **Trench method planting:** Trench method of planting is mostly used in areas that have strong winds and a long rainy season. Furrows are not made; instead trenches are made, 30 cm deep and 90 cm wide, by a trencher (tractor-operated). Fertilizers and insecticides are mixed with soil and trenches are filled with soil at a depth of 5 cm. Cane seeds are placed in the trench, 15 cm apart, or placed horizontally like stairs. In this method, more seed cane is required than in the conventional method and yield is also 10–15% higher.

Fig. 4

6. **Ring pit method planting:** This is a good method for increasing cane yield but it requires more labor at the time of planting and inter-culturing, to place two budded cane seeds in a 2,700 ring pit that has a 60 cm radius and is 30 cm deep. The pits are kept 60 cm apart. In a pit, 25–30 two-budded sets are kept in a circle. Before planting, manure, fertilizers and insecticide are mixed in soil and used to fill up to 15 cm of the pit. After the cane seed is placed in the pit, it is again filled with soil to a depth of 5 cm.

Fig. 5

Q. What is the ratio of fertilizers in a pit?

ANS. The fertilizer ratio in a pit is as under:

Urea—8 gm, Diammonium Phosphate (DAP)—20 gm, Murate of Potash (MOP)—16 gm, Zinc Sulphate—5 gm and Sulphur—5 gm.

7. **Sustainable Sugarcane Intensification (SSI) planting nursery method:** This involves raising young cane plants in a nursery using small chips taken from the cane while the balance cane is used in factory supply. The seed requirement in this method is very less. After cutting the chips of bud, a nursery can be prepared in a tray containing coco pith and vermi-compost. These seedlings must be transplanted while still young (25–35 days) with wide spacing between rows and plants (120 × 60 cm). Sufficient moisture must be provided during transplanting the seedlings. Practicing intercropping with other crops, such as soya bean, onion, green gram and kidney beans (*rajma*) allows for more effective utilization of land which also enhances soil health and biological activity in soil.

This method yields a higher number of millable cane with more weight and greater suitability for rapid multiplication of improved cane variety. Establishing a nursery bud chip plant in a tray with sand and FYM mixture, through the SSI technique, gives Co 0238 variety that attains up to twenty tillers, whereas cane from the conventional method has only five-six tillers in a clump.

Fig. 6

Q. What are the constraints of this technique?

ANS. SSI needs more attention than conventional methods. Farmers cannot just plant the crop and forget it for a month, at least not until germination is completed.

Q. What is Spaced Transplanting (STP)/Raised Bed Seedlings (RBS)?

ANS. In this method of planting, single-eyed sets are used for planting. Only 700–1,000 kg seed is sufficient for one acre of planting. Seed cost savings are 60–70%. Eye-bud to eye-bud distance is maintained at 30 cm.

Q. What is the Standard Operating Procedure (SOP) for random growth observation?

ANS. SOP for random growth observations in standing crops is as under:

Background: Growth observations are recorded to get an idea about the crop condition and the likely cane yield in the area. As yields are projected on the basis of data obtained from growth observation, plot selection for data recording and correctness of data captured is of utmost importance. This SOP aims to formulate a uniform system for data recording of growth observation parameters.

Methodology

a) Plot Selection

1. Plot Selection: Average plots should be selected for recording growth observation parameters. Average plot means that the plot should more or less represent most of the plots in the area.
2. Area for Plot Selection: Plots have to be selected from all the development circles.
3. No of Plots to be Selected: Five plots must be selected, two of early and three of the general variety, to record data from each circle.
4. Size of Plot: The selected plots must have an area of more than half an acre.

b) Data Recording

1. Selection of Rows: Three consecutive cane rows at least 20 feet inside the plot must be selected. Care has to be taken that the rows selected are from a portion of the cane field that is similar in growth to the entire field. In each row, a length of three meters must be marked in such a manner that the starting point (0) of each row is in a straight line.
2. Data to be Recorded: This includes the number of clumps and the number of millable cane in each clump in the three rows, the height of all the millable cane in one clump from each row (total three clumps) and the girth-circumference of the mother shoot at its mid-portion from all clumps of each row. For height, the cane should be measured from the ground surface to the tallest point of the cane.
3. Data recording sheet is given below.

c) Data Compilation

Data on the data recording sheet will be compiled circle wise and zone wise.

1. Data Analysis and MIS: Assuming a minimum of four circles in each zone, there will be 4 * 2 = 8 data sets for early and 4* 3 = 12 data sets for general varieties for each zone. Two data sets of minimum value for total number of millable cane and two data sets of maximum

value for total number of millable cane will be deleted to get the final data of growth observation for early variety for a zone. For general variety, three data sets of minimum and maximum value for millable cane will be deleted to get the final data for that zone.

d) **Periodicity**

1. Ratoon data will be recorded in August, September and October, and plant data will be recorded in November and December.
2. Data will be recorded in the second fortnight of the aforesaid months.

Table 2
RANDOM GROWTH OBSERVATION FORMAT—A

Name of Farmer...Observation Date ....
Father's Name.. Area (Acre).....................
Length of Row......
Zone/Circle.........Name of Kamdar (Field Asstt.)....................

S.No.	NO OF ROWS	NO. OF BUDS GERMINATED				Germination %	REMARKS
		REPLICATE I	REPLICATE II	REPLICATE III	TOTAL		
1							
2							
3							

RANDOM GROWTH OBSERVATION FORMAT—B

S.No.	NO OF ROWS	NO. OF TILLERS/ROWS/REP.				NO. OF TILLERS/HA	REMARKS
		REPLICATE I	REPLICATE II	REPLICATE III	TOTAL		
1							
2							
3							

RANDOM GROWTH OBSERVATION FORMAT-C

S.No	No of Rows	Girth & Height of cane								Total Av Girth	Total Av Height	Remarks
		Replicate I		Replicate II		Replicate III		Total				
		Girth	Height	Girth	Height	Girth	Height	Girth	Height			

SUGARCANE VARIETIES

Yield and quality depend on the adoption of a sugarcane variety, so the discussion on sugarcane varieties is critical to growers and the industry. In this chapter, we also note new cane varieties that produce maximum yield with best quality.

Q. How can sugarcane varieties be differentiated?

ANS. Sugarcane varieties are commonly differentiated by group, that is, early group and general group.

A. EARLY GROUP

Q. What are the names of some early maturing varieties in northern India?

ANS. Co 0238, Co 0118, Co 0239, Co 98014, Co 5009, CoLk 94184, CoSe 3234, CoSe 98231, CoJ 85, CoS 8436, CoS 8272 and CoS 96268.

Q. What is the parentage of some varieties of the early group?

ANS. The parentage of five early group varieties are as under:

Co 0118 = Co 8347 × Co 86011 (year of release: 2009)

Co 98014 = Co 8316 × Co 8213 (year of release: 2007)

Co 0238 = CoLk 8102 × Co 775 (year of release: 2009)

CoS 8436 = MS 68/47 × Co 1148 (year of release: 1986)

CoJ 85 = Q 63 × CoJ 70 (year of release: 2000)

Q. How can early group varieties be identified?

ANS. See the image below to understand distinguishing characteristics for some early group varieties.

Fig. 7

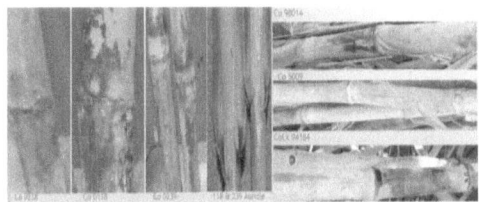

B. GENERAL GROUP

Q. What are the names of some general group varieties in northern India?

ANS. Co 5011, Co 0124, Co 86032, CoJ 88, CoS 767, CoS 7250, CoS 8279, CoS 8276, CoS 98259, CoS 97264, CoS 97261, CoS 8432, U.P. 0097, CoPb 91, CoH 167 and CoSe 1434.

Q. What is the parentage of some varieties of the general group?

ANS. The parentage of five general group varieties are as under:

Co 5011 = CoS 8436 × Co 89003 (year of release: 2014)

CoJ 88   = CoJ 82315 × Co 1148 (year of release: 2002)

CoS 7250 = CoS 8436 × Co 775 (year of release: 2009)

Co 0124 = Co 89003 (selection from ordinary cross)

CoS 767 = Co 419 × Co 313 (year of release: 1976)

Q. How can general group varieties be identified?

ANS. See the image below to understand distinguishing characteristics for some general group varieties:

Fig. 8

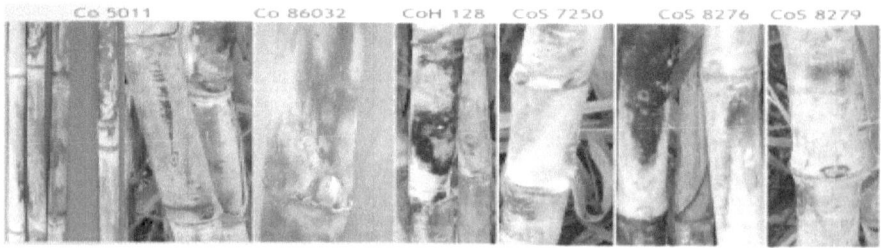

Q. What are the names of some Kenyan sugarcane varieties?

ANS. Indian sugarcane varieties such as Co 421, Co 617 and Co 945 are commonly grown in Kenya but some new varieties have also been developed by Kenya Agricultural & Livestock Research Organization-Sugarcane Research Institute (KALRO-SRI), Kisumu. These include KEN 98–367, KEN 82–601, KEN 00–3811, KEN 00–13, KEN 98–530, KEN 83–737, KEN 98–551 and KEN 00–3548. These varieties are under trial or are grown in very few areas in pockets, mostly areas covered under cane of Indian sugarcane varieties.

NUTRIENT MANAGEMENT

This chapter describes doses of fertilizers, recommendations, deficiency, how deficiency can be rectified and application time, since time plays a vital role in the fertilizer getting greater yield.

Q. Why is manure required at the time of sugarcane planting?

ANS. Manure maintains soil texture, its pH balance and its organic carbon as well as its ability to retain water. Sugarcane is a crop that lasts for a minimum of two years, so manure must be used at the time of planting. If land is being prepared for sugarcane planting, then well decomposed cow dung/FYM @ 10 tons or press mud @ 5 tons/acre must be applied a month before planting. This helps it mix with the soil completely and maintain the texture of the soil.

Q. How is green manure beneficial for sugarcane?

ANS. If we use green crops of legumes (green gram, black gram, sun hemp, soya bean, etc.) as green manure to plough the field before the pod settles or at the time of flowering, then FYM need not be used when planting.

Q. In a field prepared for sugarcane planting, if green manure has been used, what is the minimum dose of FYM to get maximum yield?

ANS. In a field where green manure has been applied, at the time of planting use only 200 kg/acre of well decomposed cow dung mixed with chemical fertilizers (SSP 150 kg + MOP 30 kg + urea 40 kg/acre) and insecticide for termite/white grub (Lesenta @ 150 gm/acre) or *buevaria bassiana* culture in the furrow before planting the seed cane. This will definitely result in maximum sugarcane yield.

Q. What is a balanced dose of fertilizers for sugarcane?

ANS. A balanced dose is usually decided with the help of soil testing but general recommendations are as under:

1. Nitrogen 180 kg/ha = 390 kg of urea
2. Phosphate 80 kg/ha = 125 kg of DAP or 375 kg of SSP

3. Potash (K)   60 kg/ha = 75 kg of MOP
4. Zinc   30 kg/ha
5. Sulphur   30 kg/ha

All fertilizers are used at the time of planting, except urea, since urea may be used in three–four equal doses. If one dose of urea is used through foliar spray at least twice with 15-day intervals 90 DaP, yield will increase.

Q. How can the quantity of fertilizers be calculated in terms of nutrient requirement?

ANS. The constant factors (values in bold below) are used to calculate the quantity of fertilizers and their nutrients percentage. Details are as under:

1. Urea (180 × **2.17** = 390 kg)
2. DAP (80 × **2.17** = 173 kg)
3. MOP (60 × **1.67** = 100 kg)
4. Calcium ammonium nitrate (? × **4.0**) (as per nitrogen recommendation × factor)
5. Single super phosphate (80 × **6.25** = 500 kg)
6. Sufla 15:15:15 (? × **6.67** =) (as per nitrogen, phosphorus, potash recommendation × factor)

Q. What is the available percentage of nutrient in fertilizer?
ANS.

Table 3: Nutrient percentage of different fertilizers.

NAME OF FERTILIZER	NUTRIENTS	PERCENTAGE
Urea	Nitrogen	46
DAP	Phosphorus	46
	Nitrogen	18
Calcium Ammonium Nitrate	Nitrogen	25
NPK (12:32:16)	Nitrogen	12
	Phosphorus	32
	Potassium	16
SSP	Phosphorus	16

	Sulphur	12
MOP (Potassium Sulphate)	Potash	50
	Sulphur	18
MOP (Potassium Chloride)	Potash	60
Ammonium Sulphate	Nitrogen	21
	Sulphur	24
Gypsum	Sulphur	13–18
	Calcium	23
Borax	Boron	11
Copper Sulphate	Copper	24
Ferrous Sulphate	Iron	19
Manganese Sulphate	Manganese	30.5
Sufla (15:15:15)	Nitrogen	15
	Phosphorus	15
	Potassium	15
Amitas (Yara)	Nitrogen	40
(France make)	Sulphur	6

Q. What is the nutrient value in different organic manures?

ANS.

Table 4: Nutrient value of different organic manures.

S.No.	Name of Organic manure	Nitrogen %	Phosphorus %	Potassium %
1	Cow dung	0.6–0.8	0.2–0.3	0.8–1.0
2	Bio compost	1.2–1.5	1.0–1.2	1.4–1.6
3	Green manure	0.5–0.7	0.2–0.3	0.8–1.6
4	Vermi-compost	2.0–2.5	2.5–2.9	1.2–1.4
5	Neem cake	5.2–5.5	1.0–1.2	1.4–1.5
6	Press mud cake	1.0–1.5	4.0–5.0	2.0–7.0
7	Bone manure	0	3.0–3.2	19.0–20.0

Q. What are nutrients?

ANS. Nutrients are essential for any plant's growth and are divided into three groups:

1. Major, primary or macro nutrients;
2. Secondary nutrients; and
3. Micro or trace nutrients.

Q. How many nutrients are included in one group?

ANS. Primary group: This includes **Nitrogen (N), Phosphorus (P)** and **Potassium (K)** which plants require in a huge quantity.

Secondary group: **Calcium (Ca), Magnesium (Mg)** and **Sulphur (S)** are required right from the beginning of plant growth but in lesser quantity than primary nutrients.

Micro/Trace nutrients: This group includes **Zinc (Zn), Iron (Fe), Copper (Cu), Manganese (Mn), Molybdenum (Mo), Boron (Bo) Chlorine (Cl) and Silica** which plants require in a limited quantity. They act as a catalyst or accelerate the enzyme activities in the metabolic process of plants.

Q. How can nutrient deficiency be identified in sugarcane crop?

ANS. Sixteen nutrients play a vital role in the growth of the crop and each nutrient deficiency reflects in crop canopy, so it can be identified by checking the crop canopy. The symptoms of deficiency in crops are shown nutrient-wise in the image below.

Fig. 9

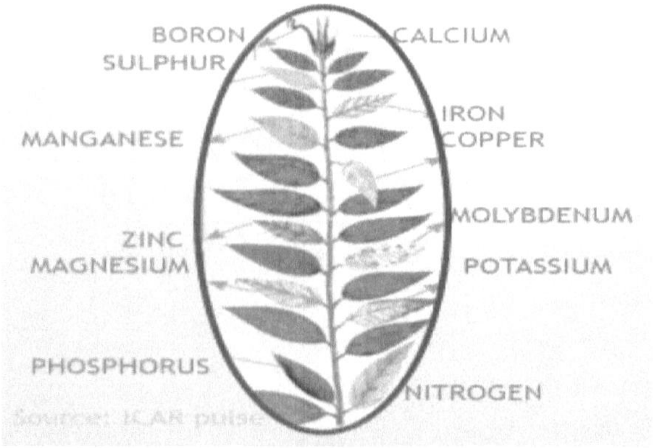

Q. What impact does acidic pH soil have on the crop?

ANS. The fertility of acidic soil suffers due to the following reasons:
1. Low availability of phosphorus (P) to the crop;
2. Toxicities of Aluminum (Al) and Mn in soil with pH lower than 5.4;
3. Deficiencies of Ca, Mg, S, Mo, Bo and Zn; and
4. Reduction of soil biological activities.

Q. How can nitrogen deficiency be identified?

ANS. If leaves turn pale yellow, plant growth is depressed, stem is thin, tillers and leaves are reduced and 90 DaP the plant leaves turn dark yellow, this is indicative of nitrogen deficiency.

Q. What are some remedial measures?

ANS. Top dressing of balanced dose and foliar application @ 2% of urea solution.

Q. How can P deficiency be identified?

ANS. Retarded leaf expansion, with leaves turning green-blue, poorly developed root system, reduced tillers and girth of the stem and older leaves reddening when the plant is young indicate P deficiency.

Q. Yellow leaves are a symptom of the deficiency of which nutrients?

ANS. Leaves turn yellow due to the deficiency of the following nutrients:

Table 5

S. No.	Nutrients	Old Leaves	Young Leaves	Character	Remarks
1	Nitrogen	Yes	No	Tips and edges dry	Internode size is reduced
2	Sulphur	No	Yes	Tips and edges dry	Stem becomes hard
3	Iron	No	Yes	First young leaves yellow	In the later stage, whole clump is whitish yellow
4	Zinc	No	Yes	With brown spots	In the later stage, the entire clump turns yellow
5	Mangenese	No	Yes	Yellow linings	Leaves and stems break in case of acute deficiency

Q. What are the optimum and critical values of nutrients in sugarcane leaves?

ANS.

Table 6: Critical nutrient values and optimum ranges for sugarcane leaves.

S.No.	Nutrients	Critical Value	Optimum range
A	Major & Macro	%	%
1	Nitrogen	1.8	2.0–2.60
2	Phosphorus	0.19	0.22–0.30
3	Potassium	0.90	1.0–1.60
4	Calcium	0.2	0.2–0.45
5	Sulphur	0.13	0.13–0.18
6	Magnesium	0.13	0.15–0.32
7	Silicon	0.50	>0.60
B	Micro	Mg/Kg	Mg/Kg
1	Zinc	15	17–32
2	Iron	50	55–105
3	Manganese	16	20–200
4	Copper	3	4–8
5	Boron	4	15–20
6	Molybdenum	0.05	-----
Source: Anderson & Bowen, 1990			

Q. What is the remedy for P deficiency?

ANS. All recommended doses must be used in basal and after that may be used in phosphate solubilizing bacteria with organic manure.

Q. What are the symptoms of K deficiency in crop?

ANS. Potassium deficiency may be associated with ratoon stunting as evidenced by decreased yield of ratoon crop when compared to plant crop. Red discoloration of the upper surface of the leaf midrib, scorching of the margins of the leaf, lodging of the crop and reduced immunity against drought are the result of the root system not developing properly.

Q. How can the deficiency of K be treated in crop?

ANS. Recommended full dose must be used as basal and foliar spray of potassium nitrate @ 1.5% solution or potassium chloride @ 10 kg/ha in 700 liters of water.

Q. Why must K be used in crop during droughts?

ANS. Potash controls the opening and closing of the stomata in the leaf, so after the application of K, the crop is not affected by drought, since photosynthesis would have been slowed down.

Q. What are the symptoms of Ca deficiency?

ANS. Due to the deficiency of Ca, leaves are thread-like and fail to expand and the spindle often becomes necrotic at the tip of older leaves. Leaves develop minute chlorotic spots and a dead center that later changes to dark reddish brown spots like rust.

Q. What are the symptoms of S deficiency?

ANS. The growth of the shoot is reduced because sugarcane requires more S for metabolism of carbohydrate. Young leaves turn yellow and the stem hardens.

Q. How can S deficiency be remedied?

ANS. Recommended full dose must be applied at the time of planting. Foliar spray of S alone or 1% solution of zinc sulphate also fulfills the requirement of S in crop. In acute deficiency of S, the symptoms can sometimes resemble those of Fe deficiency. However, in the case of S, symptoms appear on younger leaves whereas in the case of Fe, the symptoms first appear on old leaves and then the entire plant could turn whitish yellow.

Q. What are the symptoms of Mg deficiency in crop?

ANS. Most characteristics are the same as Ca deficiency, except that new leaves are light green with brownish spots. In acute deficiency, the brown spots may appear inside the stem.

Q. How can Mg deficiency be controlled?

ANS. Foliar spray of magnesium sulphate @ 10 kg/ha in 500 liters of water.

Q. How can the nutrient requirement be met?

ANS. If deficiency shows in alkaline soil then add gypsum and if in acidic soil then apply lime.

Q. What are the symptoms of Zn deficiency in sugarcane crop?

ANS. Yellowish midrib of new leaf, small size of leaf, which seems shapeless and new tillers sprouting yellow in color are the symptoms.

Q. How can the Zn requirement of crops be met?

ANS. The recommended dose must be applied during planting. To fulfill the requirement of Zn, 90 DAP 1% solution may be applied through foliar spray.

Q. What is the benefit of treating seed with zinc phosphate and sodium molibdate?

ANS. When Zn and Mo deficiency are noticed early in the crop growth stage, then seed treatment is a good method to prevent deficiencies in crop.

Q. What are the symptoms of Cu deficiency?

ANS. The deficiency of Cu is first visible on old leaves as rusty spots. In acute conditions, symptoms may appear on younger leaves too. High application rates of phosphate fertilizer may increase Cu deficiency.

Q. What are the symptoms of Fe deficiency in crop?

ANS. Symptoms include a whitish yellow lining in leaves that turns white and clumps that appear white. Iron deficiency also affects the growth of plants and the photosynthesis process.

Q. How can Fe deficiency be controlled?

ANS. Foliar spray @ 0.5–1% solution of ferrus sulphate 90 DaP or recommended dose applied at the time of planting as basal dose.

Q. How can Mn deficiency be identified?

ANS. Initially, a yellow lining develops in leaves which later turns completely yellow. In acute deficiency, the leaves and stem may break and brown spots may appear on leaves.

Q. How can the crop's Mn requirement be fulfilled?

ANS. Foliar spray may be applied 90 DaP with manganese sulphate @ 0.5% solution.

Q. What are the symptoms of Bo deficiency?

ANS. Transparent spots on leaves, wrinkling of young leaves, decreased length of internodes and stunted growth of the apical part of plants. Immature leaves may not unfurl and young plants could be brittle and bunched with many tillers when deficiency is acute.

Q. How can Bo requirement be fulfilled?

ANS. A 1% solution of boric acid/borex may be used as foliar spray 90 DaP.

Q. What are the symptoms of Mo deficiency?

ANS. Yellow lining appears in old leaves. These lines could be nearly 3 mm wide. This mostly appears in acidic soils.

Q. What are the remedial measures?

ANS. Molybdenum (50–60 gm) could be dissolved in 1,000 liters of water for one hectare and applied through foliar spray.

Q. What is the role of nitrogen in the plant?

ANS. Nitrogen is a necessary growth element that activates the use of P and potassium in the plant and is also necessary for protein formation and amino acid synthesis.

Q. What is the role of P?

ANS. Phosphorus develops the root system, photosynthesis and acceptance of minerals or food substances in plants and is also necessary to store food in plants.

Q. What is the role of K?

ANS. Potash reacts as a catalyst to increase the activities of carbohydrates and starch and the growth of plants. It is necessary for controlling the opening and closing of stomata, thus controlling photosynthesis. It is also known as the quality element of sugarcane plants.

Q. What is photosynthesis in plants?

ANS. Photosynthesis functions as a counterbalance to respiration; it takes in the carbon dioxide produced by all breathing organisms and reintroduces oxygen into the atmosphere.

Photosynthesis is written as follows:

$$6\ CO_2 + 12\ H_2O + \text{Light Energy} \rightarrow C_6H_{12}O_6 + 6\ O_2 + 6\ H_2O$$

Here, six molecules of carbon dioxide (CO_2) combine with 12 molecules of water (H_2O) using light energy. The end result is the formation of a single carbohydrate molecule ($C_6H_{12}O_6$ or glucose) along with six molecules each of breathable oxygen and water.

SOIL AND SOIL HEALTH

Without knowing soil requirements, how can yield be increased? So, we incorporated this chapter which discusses how soil can be reclaimed and soil texture maintained, what is soil pH and why soil testing is necessary?

Q. What is soil?

ANS. Soil is a natural dynamic body developed by natural forces acting on natural materials. Soil is a living medium for plant growth. It provides nutrients and water. From a farmer's point of view, soil is the portion of the earth's surface that he can plough and grow crops on to provide him food. Low soil organic carbon adversely affects the soil's physical fertility especially its water-retention capacity, root growth and proliferation.

Soil Texture

Q. How can soil health be maintained?

ANS. To maintain soil health, the following steps may be taken:

1. Adopt crop rotation.
2. Adopt intercropping system.
3. Use legume crops as green manure.
4. Use *trichoderma viride* during land preparation.
5. Use a balanced fertilizer dose.
6. Use vermi-compost @ 2 ton or press mud/filter cake @ 5 ton or FYM 10 ton/acre for a crop cycle.
7. If possible, use organic fertilizers as basal and inorganic fertilizers as foliar spray.
8. Avoid excess use of insecticides, especially red-mark level.
9. Use MB plough/Chiseler after every third year to break the hard layer of soil.

Q. Which type of soil is problematic?

ANS. Soil that has a pH level below 6.9 is called acidic; for neutral reclamation, add lime. Soil with a pH level above 7.5 is called saline (alkali); this can be reclaimed by adding gypsum.

Q. What is pH in soil?

ANS. pH is a parameter that provides information on the character of soil: whether it is acidic (<7.0 pH) or alkaline (>7.0 pH).

Q. What nutrients are available in different pH soils?

ANS. In different ranges of soil pH, nutrients are available to plants only in between a certain range (low to high). Beyond that, plant roots cannot absorb nutrients. The nutrient range for different pH soils is as under:

Fig. 10

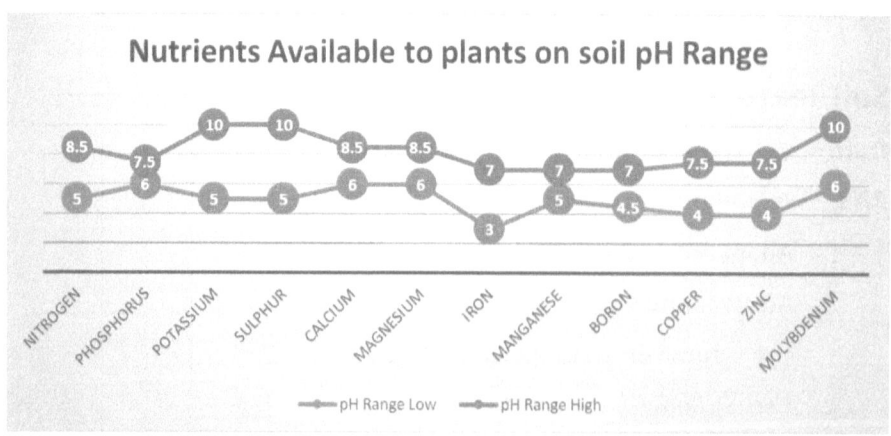

Q. Why is gypsum used in alkaline soil?

ANS. Gypsum contains 23.2% Ca and 18.6% S. When mixed in soil, it neutralizes the alkali content and improves the soil's physical properties. It is also necessary for plants to absorb NPK in soil.

Q. On the base of pH, how can soil be reclaimed?

ANS. On the basis of pH, soil can be reclaimed as under:

Table 7

Crop Status	Soil Type	pH range	Reclamation
Difficult to grow crops	Alkaline	12	Difficult to reclaim
		11	
Moderate for crop growth		10	1,500 kg/acre gypsum
		9	500–800 kg/acre gypsum
		8	
Optimum for plant growth	Neutral	7	Only recommended fertilizers
Moderately sensitive	Acidic	6	500–800 kg/acre lime
		5	
Highly sensitive		4	1,200–1,800 kg/acre lime
Difficult to grow crops		3	Difficult to reclaim
		2	

Q. What is organic carbon in soil and how is it beneficial to the crop?

ANS. Organic carbon (matter) is an important feature of soil fertility. Green manure crops or mulching the trash/residue of crops helps increase soil organic matter and control weeds. Organic carbon increases water-retention capacity in soil. If organic carbon decreases, pH balance is disturbed and ultimately crop production decreases.

Q. Does soil erode in sugarcane planting?

ANS. Sugarcane fields have relatively low levels of soil loss, due, in part, to the semi-perennial nature of sugarcane. In fact, sugarcane is typically only replanted every two or three years in India, whereas in other countries it is replanted every six or seven years.

Q. How is waterlogged soil affected?

ANS. An important chemical change that takes place when soil is waterlogged is that the pH of acid soils increases and the pH of calcareous and alkali soils decreases. In most submerged soils, manganic and ferric compounds reduce to form more soluble divalent forms that are present mainly as bicarbonates. Soil reduction under flooding is a direct consequence of exclusion of molecular oxygen, O2. Within a few hours of submergence, aerobic organisms use up trapped O2 and become quiescent or die.

Q. What is soil fertility?

ANS. The ability of soil to provide all essential plant nutrients in available form is called soil fertility.

Q. What is soil productivity?

ANS. Soil productivity is the capacity of soil to produce plants under a specified program of management.

Soil Testing

Q. Why is soil testing necessary?

ANS. Soil testing helps make a recommendation for a particular crop and determine the deficiencies, if any, so a perfect dose (balanced dose) of fertilizers can be used to achieve better yield.

Q. Why is soil mapping required in the sugar industry?

ANS. While soil classification enables the grower or manager to identify soil characteristics in the field and to provide an understanding of soil processes, it is important for users to know the location and distribution of soils on a particular farm. Cane grows well on good soils with relatively little management, but greater knowledge is required of the many poor soils in the sugar industry if they are to be conserved and managed in the best possible manner.

Q. What is the basic concept of soil mapping?

ANS. Knowledge of the range of soils, their nature and where they occur in an area is a prerequisite in any feasibility study aimed at determining the yield potential and management of sugarcane. Apart from influencing the design and management of irrigation and drainage systems, soil mapping helps in determining systems of land preparation for establishing new cane areas and other management issues dealt with in this manual related to nutrition, weed control, variety selection, the use of agricultural chemicals, pest and disease management, trash management and the optimum time to harvest.

Q. How can soil samples be collected?

ANS. To sample a field or pasture, make a map that identifies each area in the field from where sub-samples were taken. Fields or tracts of land with differences in past cropping, fertilization, liming, soil types or land use will require several composite samples. The field identification map should be used each time samples are collected. Traditionally, soil samples are collected to a depth of 9–12 inches. This depth is measured from the soil surface after non-decomposed plant materials are moved aside. This sampling depth can be significantly altered based on tillage or fertilization practices. The objective in sampling is to obtain small composite samples of soil that represent the entire area to be fertilized or limed. This composite sample comprises ten–fifteen cores or slices of soil from the sampling area. The sample must be collected from each core with a V-shaped pit as shown under:

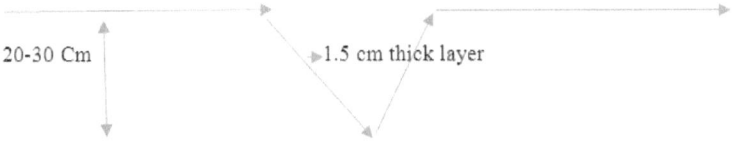

Q. Why is nitrogenous fertilizer not used before irrigation in sandy soils?

ANS. Nitrogen has the property of leaching with water and in sandy soil it goes deep down with water and then the plant cannot use the nitrogen. The purpose of using such type of fertilizer is that at least two-thirds of it is absorbed by plants.

Q. How does the soil change if we continue to grow sugarcane on it?

ANS. If organic manure/green manure is not used and sugarcane is continued to be grown, then organic carbon and nitrogen go down up to 0.6% and .011%, respectively. The depletion of organic carbon and nitrogen decreases the yield of the crop each year and increases the cost of cultivation.

Organic Farming

Q. What is organic farming?

ANS. Organic farming refers to ecological systems for producing food and fiber. Organic farming may be most widely known for what it is not; however, it is important to define what it is. Organic farming can be defined by the proactive, ecological management strategies that maintain and enhance soil fertility, prevent soil erosion, promote and enhance biological diversity and minimize the risk to human and animal health and natural resources.

Q. How can nutrition be managed in organic farming?

ANS. To produce a healthy crop, an organic farm needs to manage the soil well. This involves considering soil life, soil nutrients and soil structure. Inorganic fertilizers provide only short-term nutrients to crops and degrade soil life without adding manure. They encourage plants to grow quickly, they do not help to build good soil structure and do not improve the soil's water-retention capacity. The soil is a living system because millions of organisms are available in soil and they survive only in the presence of organic carbon/manure. These creatures are very important for recycling nutrients.

Manure or compost feeds the whole variety of life that thrives in the soil. This then turns the soil into food for plant growth. This also adds nutrients and organic matter to the soil. Green manure also provides nutrients and organic matter. It is important to remember, however, that using too much animal manure or nutrient-rich organic matter, or using it at the wrong time, could be as harmful as using inorganic fertilizers.

Q. How are organic products beneficial for humans?

ANS. Organic products benefit humans because they do not contain used pesticides, herbicides and inorganic nutrients that are used to speed up crop growth. Maintenance of organic integrity means to eliminate cross-contamination with prohibited inputs and non-certified agricultural products, the exclusion of genetically engineered organisms, synthetic fertilizers, synthetic pesticides, preventative antibiotics, growth hormones and artificial flavors, colors and preservatives.

Q. How do poly houses help in organic farming?

ANS. Poly/greenhouses help to produce organic vegetables and fruits and they helped to produce good quality of product with controlled environment and without chemical treatment. This is beneficial to the health of humans and in increasing the income of growers, who can establish poly houses with the help of governments. For technical support, growers can contact different consulting agencies that are available.

WEED MANAGEMENT

Weeds are a major problem in sugarcane crops and cost growers 30–40% yield loss every year. In this chapter, we describe mechanical as well as chemical processes that growers can easily apply to control weeds.

Q. What are weeds?

ANS. Any plant or vegetation interfering with the objective of the requirements of people: a plant growing where it is not desired or a plant whose economic value has not been discovered or a plant that is noxious, useless, unwanted or poisonous.

Q. What is the expected loss due to weeds?

ANS. Crop yield could decline 30–40% and in some cases there could be total crop failure. Weeds take away a sizeable amount of moisture and nutrients from the fields. An estimated depletion of nutrients from fertilized crop is in this fashion: N 160–165 kg, P 20–25 kg and K 200–205 kg per hectare. Weeds also provide shelter to various pests and harbor diseases.

Type of Weeds

Q. What are some sugarcane weeds?

ANS. Weeds in sugarcane crop are of two types:

Grasses or Narrow-Leaf Weeds:

Cynodon dactylon, echinochloa, echinochloa colona, paspalum conjugatum, sorghum halepense, panicum sps, setaria glauca, digitaria sanguinalis, dactyloctenium aegyptium, cyperus rotundus, poa annua and leptochloa chinensis

Broad-Leaf Weeds

Chenopodium album, convolvulus arvensis L., amaranthus spp. L., portulaca oleraceae L., parthenium historica, commelina bengalensis L., trianthema portulacastrum L. digera arvensis, eclipta alba, casueria axilaris, ageratum conejoides and ipomoe sps.

Q. How can these weeds be identified?

ANS. Most common weeds in sugarcane crop are as under:

Fig. 11

Q. What is striga?

ANS. Striga is a very dangerous weed in sugarcane and corn crops, especially in African countries.

Q. How can this weed be controlled?

ANS. Before the striga plants start to flower, hoeing and hand weeding must be undertaken. Late weeding requires the burning of collected plants to kill striga seeds. Never put them into the compost pit. Green gram beans, soya beans and other legumes can be used for intercropping to improve the fertility of soil and cover the empty spaces in sugarcane crop and deprive the parasitic weeds of favored host plants.

Q. How do weeds affect sugarcane?

ANS. Doob grass (*cynodon dactylon*) and the cogon grass (*imperata cylindrica*) are known to play hosts to the ratoon stunting disease of sugarcane. Thus, weeds essentially harm young sugarcane sprouts by depriving them of moisture, nutrients and sunlight. Poor growth of cane resulting from weed infestation also affects quality. Weeds present in the furrows, that is, along the cane rows, cause more harm than those present in the inter-row spaces during early crop growth sub-periods. Thus, the initial 90–120 days of crop growth is considered most critical for weed competition.

Q. How beneficial is de-trashing/mulching in weed control?

ANS. Retention of such relatively large amounts of trash residue in high rainfall and irrigated areas provides an opportunity for recycling of organic matter and nutrients and carbon sequestration in the soil. De-trashing/mulching have the following benefits:

1. Control of weeds since they are not allowed to grow.
2. Reduction of infestation of pests such as scale insects, internode borer, mealy bug, etc.
3. Retention of soil moisture so plants can easily obtain nutrients due to less competition.
4. Increase in the thickness of the stem.
5. Increase in the cane yield, up to 10%, if de-trashing is undertaken when the cane is six–eight months, in irrigated areas, and nine–ten months in rain-fed areas.

Q. How can organic weedicide be prepared?

ANS. Collect 10 liters of cow urine in a plastic pot and mix in 2 kg salt, 100 ml neem oil and the juice of one lemon. Properly stir this mixture for up to 15 minutes. The solution is then ready for use as organic herbicide.

Q. What is the dosage of organic herbicide to be applied?

ANS. Organic herbicide dose is 80 liters per acre.

Q. What precautions can be taken?

ANS. Spraying should be undertaken in the morning, when the field is dry or when there is no water stagnation.

Q. What are the benefits of organic herbicide?

ANS. The benefits of organic herbicide are:

1. It is very cheap and easy to prepare.
2. Its components are easily available.
3. It is very effective against weeds; results are visible after two days.
4. It also controls insect pests.
5. It is risk free.

Q. How can weeds be controlled in sugarcane crop?

ANS. Weed control in sugarcane is done by adopting mechanical or chemical methods.

A) Mechanical Control

1. Hoeing with the help of a spade or inter-culturing with the help of power tiller/rotavator.
2. Hoeing with the help of three-tine cultivator followed by manual weed removal from crop rows.
3. The process must be repeated frequently 30 DaP.
4. This method not only removes weeds but also increases sprouting tillers and destroys soil insects and enhances aeration in the soil.
5. The use of weeder suppresses/controls the growth of weeds and enhances the development of the root system of the cane crop due to aeration in soil.

6. Power tiller/power rotavator can be used as a weeder.
7. A laborer with power tiller can complete intercultural operation in one-two acres in a day.
8. The intercultural practice may be used a minimum of three times.
9. First weeding can be undertaken 30 DaP (blind weeding).
10. The third weeding can be with the second dose of urea. Thereafter, earthing up may be completed.

B) Chemical Control

1. Pre-emergence of the weeds in plant and ratoon crops, spray Merlin @ 75 gm + Sencor @ 400 gm/acre in 300 liters of water through a boom sprayer (Fig.12) or
2. 1–3 DaP spray Roundup @ 1 liters/acre in 300 liters of water during pre-emergence of sugarcane crop with sufficient moisture in the upper layer of soil or
3. 1–3 DaP spray Atrazine @ 400 gm + Sulfentrazone @ 350 gm/acre to control weeds.
4. 1–3 DaP spray Sencor (Metrabuzine) @ 400 gm + Sunrice (Ethoxysulfuron) @ 150 gm/acre in 300 liters of water to destroy all narrow and broad leaf weeds.
5. 30–40 DaP spray Laudis @ 120 gm + Atrazine @ 400 gm/acre in 300 liters of water to control broad and narrow leaf weeds.
6. 45–60 DaP spray Sempra @ 36 gm/acre in wet conditions to control narrow leaf grasses, especially Dactyon (*Motha/Dilla*).
7. 30 DaP spray 2-4-D (Durex) @ 400 gm/acre in 300 liters of water to destroy all broad leaf weeds.

FOLIAR SPRAY ON SUGARCANE

Foliar spray plays a very important role in ratoon crop because these are frequently attacked by some sap-sucking insect pests that can damage the crop. If the grower applies a foliar spray of nitrogenous fertilizer with insecticide, it benefits the crop and yield increases.

Q. Why should foliar spray of urea be used in sugarcane?

ANS. When nitrogenous fertilizer is applied through basal dose, only 38–46% is absorbed by the crop, whereas when applied though foliar spray 100% is absorbed by the plants within a few hours and no extra dose goes to the weeds.

Q. When and how should fertilizer be used through foliar application?

ANS. Generally, insecticide through spray is used on sugarcane 90 DaP, in a solution of urea @ 20 kg/ha in 800 liters of water.

Fig. 12

Q. Can foliar spray be applied without irrigating the crop?

ANS. Foliar spray of urea cannot be used in drought conditions. If it is applied in a dry field, the field must immediately be irrigated otherwise the sugarcane crop may be damaged (burning effect).

Q. How does foliar application of urea benefit sugarcane crop?

ANS. Foliar application ensures that 90% of nitrogen enters the leaf through stomata and participates directly in the metabolic reaction of the plant and that maximum quantity reaches the final end without wastage of time. Thus, results are fast and good.

Q. What is the quantity of urea to be used through foliar application?

ANS. In sugarcane, 90 DaP foliar spray can be used twice with 15-day intervals and 20 kg urea/ha at one time. This is equal to 50 kg/ha of nitrogen used four times.

Q. Can NPK be used in foliar spray?

ANS. Yes, NPK is more beneficial to crop growth and fulfills plants' requirement instantly. The solution of NPK should be used @ 1 kg/acre in 250 liters of water. NPK used through foliar spray also helps plants bear drought and regularizes plant growth.

Q. How can yield be increased by using OLF in foliar spray?

ANS. When Organic Liquid Fertilizer (OLF) is used in foliar spray @ 300 ml/tank (20 liters) at least four times, it increases the yield of the sugarcane crop by 15–20% without increasing the dose of fertilizers.

Q. What is OLF?

ANS. OLF, an organic product, has the potential to promote growth and provide immunity to plant systems. It consists of nine products: water, cow dung, cow urine, milk, curd, jaggery, ghee, banana and coconut water. When properly mixed and used, these have miraculous effects.

Q. How can organic liquid fertilizer (OLF) be prepared and what are its ingredients?

ANS. All ingredients and quantities are as under:
- Cow dung: 7 kg
- Cow ghee: 1 kg
- Cow urine: 10 liters
- Cow milk: 3 liters
- Cow curd: 2 liters
- Water: 10 liters
- Coconut water: 3 liters

- Sugarcane juice: 3 kg or Jaggery: 500 gm
- Well-ripened banana: 12

Method of preparation: In a wide-mouthed earthenware pot, concrete tank or plastic can, placed in the shade, mix cow dung and cow ghee thoroughly, both in the morning and evening and keep it for three days. After three days, mix cow urine and water and keep the mixture for fifteen days, regularly mixing it in the morning and evening. After fifteen days, mix the remaining ingredients and the OLF will be ready after thirty days. (Care should be taken not to mix buffalo products. The products of local breeds of cow are said to have greater potency than exotic breeds.) The mixture should be kept in the shade and covered with a wire mesh or plastic mosquito net to prevent houseflies from laying eggs and the formation of maggots in the solution. If sugarcane juice is not available, add 500 gm of jaggery dissolved in 3 liters of water.

Q. What are the physicochemical properties of OLF?

ANS. The physicochemical and biological properties of OLF are as under:

Table 8

Chemical composition	
pH	5.45
EC dSm2	10.22
Total N (ppm)	229
Total P (ppm)	209
Total K (ppm)	232
Indol Acitic Acid (IAA) (ppm)	8.5
Gibberelic Acid (GA) (ppm)	3.5

Note: ppm = parts per million

Source: TNAU –Organic Farming (2012)

The physicochemical properties of OLF reveal that they possess almost all the major nutrients, micro nutrients and growth hormones (IAA and GA) required for crop growth.

Q. What dose of OLF should be used for sugarcane?

ANS. The recommended dose of OLF for different types of application is:

Foliar Spray Application

Investigations showed that a 3% solution was most effective compared to higher and lower concentrations. Three liters of OLF for every 100 liters of water is ideal for all crops. Power sprayers of 10-liter capacity may need 300 ml/tank. When sprayed with power sprayer, sediments are to be filtered and when sprayed with hand-operated sprayers, a nozzle with a higher pore size must be used.

Through Irrigation

The OLF solution can be mixed with irrigation water at 50 liters per hectare either through drip or flow irrigation.

Seed/Seedling Treatment

A 3% solution of OLF can be used to soak the seeds or to dip the seedlings in before planting. Rhizomes of turmeric, ginger and sets of sugarcane can be soaked for thirty minutes before planting.

Q. What impact does OLF have on sugarcane crop?

ANS. The following impact can be seen after foliar application of OLF:

Yield: Under normal circumstances, when land is converted to organic farming from inorganic systems of culture, yield depression takes place. The key feature of OLF is its efficacy in restoring the yield of all crops when the land is converted from an inorganic culture to organic culture from the very first year. The harvest is advanced by 15 days in all crops. It not only enhances the shelf life of vegetables, fruits and grains but also improves the taste. By reducing or replacing costly chemical inputs, OLF ensures higher profit and liberates organic farmers from loan.

Drought Hardiness: OLF creates a thin oily film on leaves and stems, thus reducing the evaporation of water. The deep and extensive roots developed by the plants can withstand long dry periods. Both these factors contribute to reduce irrigation water requirement by 30% and ensure drought hardiness.

Q. What is the time schedule for OLF application?

ANS.

Table 9: Time of OLF application on different crops.

Crops	Time schedule
Sugarcane	**90, 110, 125 and 140 days after planting**
Rice	10, 15, 30 and 50 days after transplanting
Sunflower	30, 45 and 60 days after sowing
Okra	30, 45, 60 and 75 days after sowing
Tomato	Nursery and 40 days after transplanting; seed treatment with 1% solution for 12 hours
Onion	45 and 60 days after transplanting

Source: TNAU-Organic Farming (2012)

Q. Can OLF be used for human diseases?

ANS. OLF is also effective for all these human diseases: **AIDS/HIV, psoriasis, neurological disorders,** *diabetes mellitus*, **pulmonary tuberculosis and arthritis.** The dosage should be 50 ml of filtered OLF mixed with 200 ml of water, tender coconut water or fruit juice and taken orally on an empty stomach in the morning.

DISEASES OF SUGARCANE

Many diseases have been reported in the crop worldwide. If sugarcane professionals have up-to-date knowledge about the time of infestation and how diseases can be identified, then they can educate growers to ensure better crop yield and protection from such diseases.

Q. How many diseases can affect the sugarcane crop?

ANS. About sixty diseases have been known to infect the sugarcane crop. However, some affect only the plant and some only the ratoon crop. Main diseases include red rot, wilt, smut, top rot, leaf scale, pokkah boeing, grassy shoot and ratoon stunting.

Q. Can these diseases be shown on a monthly calendar?

ANS. Yes.

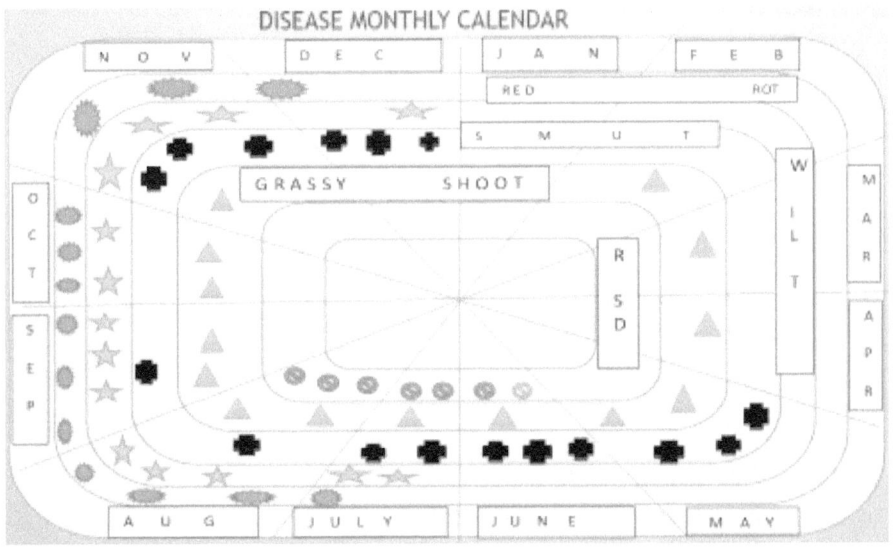

Fig. 13: The main diseases affecting sugarcane, plotted on a monthly calendar.

Q. What is the SOP for disease observations in the field?

ANS. The SOP for disease observations are as under:
1. Collect disease incidence data from each circle of each zone.
2. Select a representative block of twenty plots (preferably, ten ratoon plots and ten plant plots) for surveillance.
3. Record disease incidence from all the twenty plots; separate data must be collected for the ratoon and plant crop—data recording sheet for both crops should be separate.
4. Record reading at least 5 meters deep from the border of the crop.
5. Select three consecutive rows of 10-meter length from a block.
6. Mark the block with red ribbon for cross-checking.
7. Take observations for disease simultaneously.
8. Consider number of clumps to calculate incidence percentage for all diseases.

Table 10: Economic Threshold Levels.

Sl	Pest/ Disease	Basis of calculation	Incidence Level		
			Below ETL	Above ETL	Serious
1	Red Rot	No. of Clumps	<5%	5–10%	Above 10%
2	Smut	No. of Clumps	<5%	5–10%	Above 10%
3	Wilt	No. of Clumps	<5%	5–10%	Above 10%
4	Grassy Shoot Disease	No. of Clumps	<5%	5–10%	Above 10%

Procedure for selecting the rows and plants to collect the data:
1. Select representative rows on a random basis.
2. Consider number of clumps for disease.
3. Take visual area of the plot.
4. Add up the area of all the plots of the selected block to calculate area of the observation block.
5. Add up area of all the affected plots in the block to calculate area affected.

6. Separately take area of circle, area of ratoon and area of plant cane.
7. Use separate data sheets for ratoon and plant crops.

Table 11

Row number	No. of clumps	Affected Clumps			
		Red Rot	Smut	Grassy Shoot Disease	Wilt
Row 1					
Row 2					
Row 3					
Total %					
Level of Incidence					
Area of the Observation Block (Ha)					
Area Affected (Ha)					
Area of the Circle (Ha)					
Area Affected in the Circle (after Extrapolation) (Ha)					

Q. What are the symptoms of each disease that mainly affect sugarcane crop?

ANS. About sixty diseases have been known to infect sugarcane but these mainly affect crop survival and productivity. The symptoms and control measures of different diseases are discussed individually below.

Q. What is Red Rot in sugarcane?

ANS. Red Rot is a very serious disease in sugarcane and it can destroy the cane variety. This is a fungal disease caused by *colletotrichum falcatum*. Also known as cancer of sugarcane, it can spread through seed, water and soil. It is responsible for 2.5–5.0% sugar loss in the last phase; however, in case of acute infestation, the entire crop could be damaged.

Q. What are the symptoms of Red Rot in sugarcane?

ANS. Details of symptoms are as under:

Fig. 14

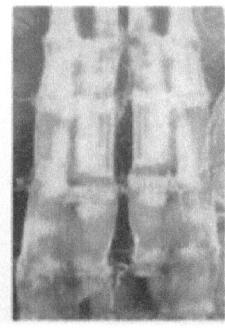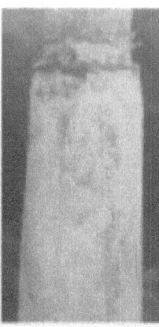

- The third and fourth leaves of the affected plant first turn yellow with red spots on the midrib. Then, these leaves dry up.
- When the cane is cut longitudinally, red spots interspersed with white spots appear.
- When the cane is cut, it gives off a very offensive smell, like vinegar.
- In cases of acute infestation, the spore of the fungus appears outside the internode.

Q. How can this disease be controlled?

ANS. This disease is very dangerous for sugarcane and an infested field cannot be cured perfectly; only precautions can be made to check that the disease has not transferred to other fields.

Control Measures:
1. No effective control has been found for this disease.
2. Precaution must be taken at the time of seed selection and planting.
3. Crop rotation must be implemented in the field.
4. Effected clumps must be dug up and bleaching powder or 1% solution of Carbendezim/Rexil must be used on the stump. The spread of the spores of fungus must be checked with water or soil.

Q. When was Red Rot discovered in India?

ANS. Red Rot was first reported in 1893 by Went. It was called *laal kandua*. Thereafter, it was reported in 1901 by Barber and in 1906 Butler gave it the name *laal sadan*.

Q. What is Wilt in sugarcane?

ANS. This disease is also caused by a fungus, namely *cephalosporium sacchari* and *fusarium moniliforme*. The spore of this fungus can enter the cane through any pore and can be carried by termites, borers, etc. A main carrier of this disease is root borer. Affected cane will not be able to germinate. This disease mainly appears after the monsoon (from July).

Q. What are the symptoms of Wilt?

ANS. Wilt in sugarcane can be identified from the following symptoms:

Fig. 15

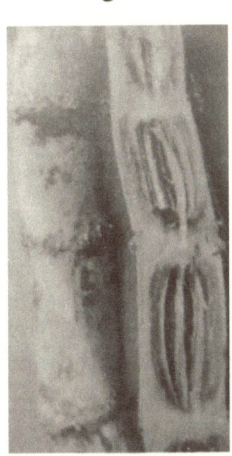

1. First symptoms appear on apical leaves which turn yellow.
2. In the secondary stage, these leaves could dry up.
3. A white thread-like substance is visible when the cane is cut in between internodes.
4. Internodes turn brown.
5. Internodes shrink.

Q. How can this disease be controlled?

ANS. To control this disease, the following steps may be adopted:

1. Soil should be treated with *trichoderma viride* at the time of land preparation.
2. Seed must be treated with fungicide and insecticide (Rexil + Gaucho).

3. In June, the last 10–12 quintal/ha of neem cake must be used and the earthing up technique applied.
4. In August, Quinalphos 25 EC @ 5 liters/ha should be used through drenching. The field should be irrigated immediately to control root borer, which is a chief cause of this disease.
5. To reduce infestation, onion, garlic and coriander should be used for intercropping.
6. Rice–sugarcane crop rotation must be used.

Q. What is the difference between Red Rot and Wilt and how can these be identified in standing crop?

ANS. Red Rot mostly affects waterlogged areas whereas wilt is seen in dry land. On breaking, the stem of Red Rot affected cane emits an offensive smell although no physical change is visible on the outside of the stem. Wilt-affected stems do not break and when they are cut, there is no smell although the internode shrinks.

Q. What is Smut?

ANS. This disease appears mainly in ratoon crops. It is a fungal disease caused by *ustilago centenary*. A two–four month-old infected plant shows a black whip-like structure with millions of spores covered with a transparent membrane. The plant has a thin stalk and is often stunted. More tillers than normal result in leaves that are more slender and weak. In the first stage of infestation, the distance between the leaves increases.

Q. What are the symptoms of Smut?

ANS. Smut in sugarcane can be identified by the following symptoms:

Fig. 16

1. The top of the plant would be black and whip-like, generally in a U-shape.
2. Primary infestation appears in older plants whereas secondary infestation can appear in younger plants.

Control:

1. Precaution must be employed in seed selection.
2. Seed cane must be treated in an MHAT plant at 54 °C for one hour and then treated with Baleton @ 0.1% solution.
3. Rexil 50 ml + Gaucho FS 600 @ 200ml/acre in 400 liters of water must be used for seed treatment.

Q. What is Pokkah Boeing in sugarcane?
ANS. This is also caused by fungus and its symptoms are as under:

Fig. 17

1. The symptoms of this disease appear between July and September.
2. Apical leaves can appear tangled.
3. The top of the cane lacks leaves and is like the spine. Thus, the affected plant looks like it is missing the top and becomes thin.
4. The growth of the plant eventually stops.

Control:

1. Foliar spray of Copper Oxichloride @ 2 gm/liter of water can be used.
2. Foliar spray of Carbandezim/Rexil @ 1 gm/liter of water can be used.

Q. What is Top Rot in sugarcane?

ANS. This disease also appears during the rainy season and is caused by *fusarium moniliforme*. The stem of the plant appears to be healthy whereas the top of the cane is damaged and eventually dries up. At the end of the rainy season, the plant can be healthy. This disease appears only in select varieties.

Q. What is Leaf Scale in sugarcane?

ANS. This disease appears in September–October. All the leaves are damaged and dry up and all eye buds sprout. Due to this, the cane cannot be used in seed.

Q. What is Grassy Shoot Disease (GSD) in sugarcane?

ANS. GSD is now common. It mainly appears in old varieties but due to phytoplasma it can spread to new varieties as well. If a cutter is used on infected crop and then used on fresh (unaffected) cane, this disease could appear in new varieties too.

Q. What are the symptoms of GSD?

ANS. The disease is more severe in ratoon rather than plant crops. It reduces juice quality, plant height and cane yield drastically. It is characterized by the proliferation of vegetative buds from the base of the cane, giving rise to a crowded bunch of tillers bearing narrow leaves. The tillers bear pale yellow to completely chlorotic (Fig.18) leaves. A diseased clump produces one or two thin, weak and small canes. In the plant crop, young leaves of diseased plants are white and the buds at nodes germinate from top downwards and are also white and hence called "albino." The disease primarily spreads through planting material and within the crop through insect vectors.

Fig. 18

Q. How can GSD be controlled?

ANS. Precautions must be taken in seed selection and seed treatment at the time of planting:

1. Seed must be selected from certified/verified nurseries.
2. Seed must be treated with Rexil 50 ml + Gaucho 300 ml/acre.
3. Seed must be treated in a MHAT plant for two hours at 54 °C.
4. After infestation, the infested clump must be stumped out from the field.
5. All stumped out clumps must be burnt.
6. The spade/implement used in stumping the clump must not be used in a healthy field.

Q. What is ring spot on leaves?

ANS. *Helminthosporium sacchari* is the scientific name of eye spot or ring spot disease. Thick canes, particularly older leaves, are more severely attacked in rainy weather. Disease symptoms first appear on the foliage as dark green oval or spherical spots, later developing straw color, from August. When the plant grows in size, the central portion dies and the straw colored tissue is surrounded by a thin red-brown band. When severe, the leaves collapse and dry prematurely. In the central straw-colored portion, many pinhead sized fruiting bodies develop in concentric rings. Juice quality is affected. Cool weather usually favors the disease.

Q. How can eye spot disease be controlled?

ANS. Foliar spray of Folicur @ 300 ml/acre in 400 liters of water can be applied.

Q. What is rust in sugarcane?

ANS. Sugarcane can be infected by brown rust (*puccinia melanocephala*) and orange rust (*puccinia kuehnii*). Brown rust is the more common pathogen and occurs worldwide.

Q. What are the control measures that can be adopted to deal with rust?

ANS. In commercial practice, rust is controlled only by planting resistant varieties.

Q. What is pineapple disease?

ANS. The scientific name of the fungus is *ceratocystis paradoxa*. It primarily affects sugarcane setts. Diseased setts when planted may rot to a height of 15–25 cm and turn chlorotic. Eventually, the leaves may wither and the shoots wilt. The central portion of the affected shoot turns red and rots. A pineapple smell is associated with the rotting and hence the name. The fungus is soil borne and enters the setts through cut ends.

Q. How can an affected field be identified?

ANS. Setts affected by pineapple disease may decay before buds germinate or young shoots may die shortly after emergence. Disease can result in crops having a patchy, uneven appearance. When severe, the disease may seriously reduce germination.

Q. Why do affected cane give off the odor of pineapple?

ANS. The odor is due to the formation of ethyl acetate caused by metabolic changes.

Q. How can pineapple disease be controlled in sugarcane?

ANS. Resistant varieties must be used and setts should be treated with fungicide (Rexil or Bavistin) for 10–15 minutes.

INSECT PEST MANAGEMENT

Sugarcane seems to be particularly at risk from insect pest infestation. In India, 208 types of insect pests have been reported in crops, of which one dozen are extremely harmful and could reduce yield. Sugarcane professionals must be aware about time of infestation, control measures and whether chemicals or bio methods are appropriate to deal with the same.

Q. How many insect pests infest sugarcane crops?

ANS. Sugarcane crops can be infested by 208 insect pests but about one dozen cause considerable damage in Uttar Pradesh (India).

Q. What is economic threshold and economic injury level?

ANS. Economic Threshold (ET) is defined as, "the population density at which control action should be initiated to prevent an increasing pest population (injury) from reaching the economic injury level (EIL), which is the lowest population density that will cause economic damage." Although usually measured in insect density, the ET is actually a time to take action, that is, numbers are simply an index of that time.

Q. What is the SOP for a pest surveillance report?

ANS.

- Collect pest incidence data from each circle of each zone.
- Select a representative block of twenty plots (preferably ten ratoon and ten plant plots) for surveillance.
- Record pest incidence from all twenty plots with separate data for the ratoon and plant crops; data recording sheets for both crops should be separate
- Record reading at least 5-meter deep from the border of the crop.
- Select three consecutive rows of 10-meter length to form a block.
- Mark the block with red ribbon for cross-checking.

- Consider number of shoots to calculate incidence levels of pests such as early shoot borer, top borer, root borer, thrips, etc.
- Consider number of clumps to calculate incidence levels for pests such as termite, white grub, black bug, mealy bug, etc.

Table 12: ETL Levels.

Sl	Pest	Basis of calculation	Incidence Level		
			Below ETL	Above ETL	Serious
1	Termite	No. of Clumps	<5%	5–10%	Above 10%
2	White Grub	No. of Clumps	<5%	5–10%	Above 10%
3	Black Bug	No. of Clumps	<15%	15–22%	Above 22%
4	Mealy Bug	No. of Clumps	<5%	5–10%	Above 10%
5	Early Shoot Borer	No. of Shoots	<15%	15–22%	Above 22%
6	Top Borer	No. of Shoots	<10%	10–15%	Above 15%
7	Root Borer	No. of Shoots	<10%	10–15%	Above 15%
8	Thrips	No. of Shoots	<15%	15–22%	Above 22%

Procedure for preparing ETL

Select representative rows on a random basis.

Take visual area of the plot.

Add up the area of all the plots of the selected block to calculate area of the observation block.

Add up the area of all the affected plots in the block to calculate area affected.

Separately take area of circle, area of ratoon and area of plant cane.

Use separate data sheets for ratoon and plant crops.

Pests

Table 13

Row no.	No. of clumps	No. of Shoots	Affected Clumps				No. of Affected Shoots				
			Termite	White Grub	Black Bug	Mealy Bug	Early Shoot Borer	Top Borer	Root Borer	Thrips	Any other
Row 1											
Row 2											
Row 3											
Total/%											
Level of Incidence											
Area of the Observation Block (Ha)											
Area Affected (Ha)											
Area of the Circle (Ha)											
Area Affected in the Circle (after Extrapolation) (Ha)											

Q. Can these insect pest infestations be shown on a monthly calendar?

ANS. Yes.

Fig.19: Annual calendar of insect pest infestations.

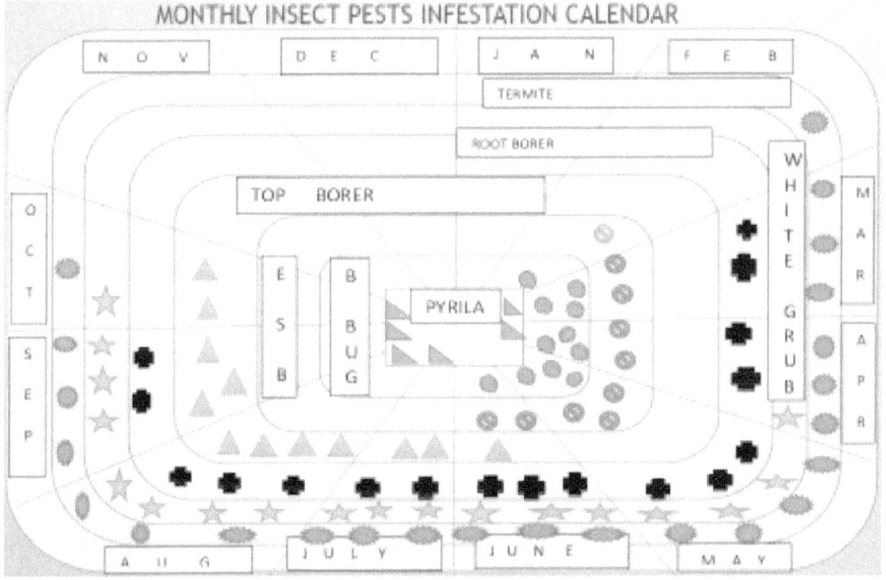

Q. What is *panch kadha* (five-plant extract)?

ANS. In an earthenware pot, collect the leaves or fruits of neem, *aakh (madaar), mahu, dhatura* and *sitaphal*. Mix 2 kg of each with 9–10 liters of cow urine and 12–15 liters of water. This mixture can be boiled on a slow heat until all the leaves/fruits are dissolved in the solution. After filtering the solution, use @ 150–200 ml/15 liters of water against any insect pest. This solution is called *panch kadha* (five-plant extract).

Q. What are termites? What kind of damage can they cause sugarcane?

ANS. These are social insects that live in colonies and occur in different morphological forms as workers and soldiers and sexual forms (king and queen) (Fig.20). After monsoon, winged sexual forms appear. After mating, the queen loses its wings, re-enters the soil, increases in size and starts laying eggs continuously for seven–ten years.

When termites attack sugarcane, canes get filled up with mud galleries after internal tissue is eaten up. Old leaves of infested plants begin to dry first and severely damaged plants can then easily be pulled out. Germinating setts are hollowed out and plant population is drastically reduced. If termite infestation is not detected and controlled in time, the entire crop could be destroyed.

Fig. 20

Worker   Queen   Damaged cane

Q. How can termite in sugarcane be controlled?

ANS. The following termite control measures could be adopted:
1. The cane seed could be treated with Confidor (imidacloprid) @ 125 ml/acre in 125 liters of water.
2. At the time of planting, the seed in the furrow should be drenched with Lesenta @ 150 gm/acre in 400 liters of water.
3. At the time of planting, the seed in the furrow could be treated/drenched in Dontatsu @ 150 gm/acre in 400 liters of water.
4. At the time of planting, the seed in the furrow can be drenched in the mixture of cow urine (details in other questions).

Q. Can termites be controlled using organic methods?

ANS. Yes, to control termites, dig a pit 1 × 1 feet deep and wide along the border line and in the corners of the field. Fill this with raw cow dung for one week. A week later, all the termites will have accumulated in the cow dung within the pit. The dung should be drawn out and either put in pond/water or burnt with kerosene oil. This activity can be repeated three to four times within the crop cycle.

Q. How can neem be used to control termites?

ANS. We can control termite and white grub with the application of neem in the following manner:

1. Neem leaves: In an earthenware pot, collect 5 liters of cow urine and 2 kg of neem leaves and shut it tight with a lid for 10–15 days. After filtering, use this solution @ 250–300 ml per 15 liters of water against soil insects, sap-sucking insects and leaf-cutting insects on all crops/vegetables.

2. Neem cake: Mix 100 kg of well-decomposed cow dung and 100 kg of neem cake properly and use in furrows at the time of planting. This helps control termites, white grub and harmful nematodes in the soil.

3. Neem seed (*neemboli*): In an earthenware pot, take 2 kg seed powder and properly mix it with 5–7 liters of cow urine and 10–15 liters of water and boil it so that half the mixture remains. This mixture is ready for use @ 100–150 ml per 15 liters of water against sap-sucking and leaf-cutting insect pests.

Q. What species of termites typically infest the sugarcane crop in Uttar Pradesh?

ANS. Five species of termites have been identified:

1. *Odontotermes obesus*
2. *Odontotermes ajumathi*
3. *Odontotermes taprobens*
4. *Odontotermes bengaloriencis*
5. *Odontotermes paratoxelis*.

Q. In which stage, can termites do maximum damage to the crop?

ANS. If seed treatment has not been proper, then the termite infestation could start at planting time. This is an important stage because germination of any crop is critical to its production.

Q. What is white grub?

ANS. White grub is a very serious pest in sugarcane, especially in western Uttar Pradesh.

Q. Which species of white grub can cause damage to sugarcane crops?

ANS. Six species of white grub have caused damage to the sugarcane crop in Uttar Pradesh. Two main species are *holotrichia consanguinea* and *holotrichia serrata*.

Q. How does white grub damage the crop?

ANS. White grub adults have reddish brown elytra. Grubs are fleshy, have a yellow head and a C-shaped body. The wrinkled part of the body is dirty white and the posterior smooth portion is brownish. They feed on sugarcane roots, one generation per year (Fig.21). Adult white grub hibernate in soil till the monsoon showers, when they emerge and feed on trees and return to the soil to lay eggs. Grubs feeding on the roots and rootlets of sugarcane cause the crop to start drying up and large patches of the crop wither. Adults feed on perennial trees (*neem, ber, jamun,* guava and *sheesham*) soon after the monsoon showers.

Fig. 21

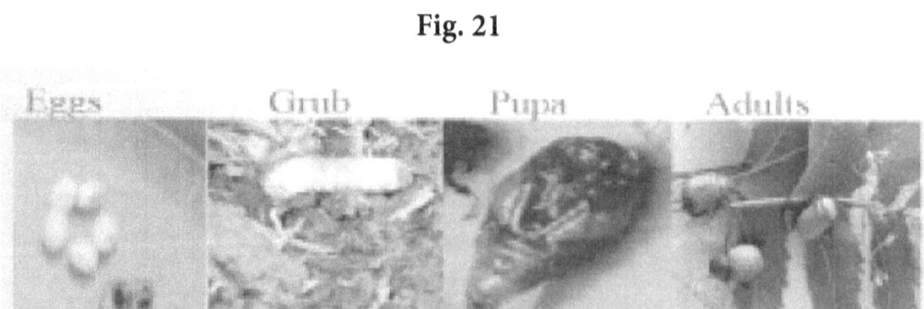

Q. White grub first time in India where was noticed?

ANS. In India, white grub was first described in 1956 in Dalmia Nagar (Bihar).

Q. Do white grub have a complete metamorphosis?

ANS. Yes, it has complete metamorphosis. The life cycle of white grub is divided into four stages:

1. Egg (8–10 days)
2. Larva (60–90 days)
3. Pupa (14–21 days)
4. Adult (180–200 days).

Q. When and where do female white grub lay eggs?

ANS. Adults appear in May–June after the first shower. They then mate on perennial trees and consume the leaves of such trees from evening to morning. The female then goes off to lay eggs in sandy soil on the border of the field at 5–10 cm depth.

Q. How can white grub be controlled?

ANS. After the first shower in May–June, foliar spray of Decis/Roket/Monocrotophos can be used on the host tree of adult. Thereafter, all adults can be collected from the ground under the host tree and burnt. If some adults are left, then 15–20 days after first shower, Chlorpyriphos/Phorate/Furadan/Quinalphos can be used @ 30 kg/ha in wet fields. The caterpillar of this pest can be controlled only at the first instar of larva. Fifteen–twenty days after the first shower, this first instar can be found in the upper layer (5–10 cm depth) of the soil.

Q. What is the scientific name of root borer?

ANS. The scientific name of root borer is *emmalocera depressella*.

Q. What kind of damage does it do to the crop?

ANS. The damage is very peculiar because the caterpillars enter at the base of stem, very close to the surface of soil. The larvae do not move upward but remain at the base. The "dead hearts" do not emit any offensive smell and cannot be easily pulled out.

Q. In which month is root borer infestation noticed?

ANS. Root borer infestation has been noticed from June to September.

Q. Which crop reduces root borer infestation?

ANS. If red gram is grown on the boundaries of the sugarcane field, it reduces or checks the infestation of root borer, because red gram is the first host crop and a favorite of root borer adults. The adult root borer lays eggs only on red gram, thus reducing the root borer attack on sugarcane.

Q. How can root borers be controlled?

ANS. At the end of May or beginning of June, Quinalphos @ 2 liter/acre can be made into a solution in 5 liters of water and then mixed in 30–40 kg sand and broadcast in the furrow. The field must then be irrigated immediately. The same application should be repeated at the end of July.

Q. What is early shoot borer (ESB)?

ANS. Its scientific name is *chilo infuscatelus*. Its eggs are straw colored and scale-like (overlapping), usually laid on the leaf sheath or on the underside of the leaf blade. Larvae are slender, white and have five violet stripes along the body with a dark brown head. Larvae move upward and downward as far as the roots, thus cutting off the central leaf spindle causing a dead heart, which can be easily pulled out and which emits an offensive smell. The central leaf sheath within the stem gets rotten and when pulled out gives off a very offensive smell. The most characteristic feature of the dead heart is that the central leaf sheath dries up while the other leaves remain green for a longer time.

Q. How can ESB be controlled without insecticides?

ANS. If the field is irrigated frequently and not allowed to dry, then the infestation is minimized. If onion, garlic or coriander is used as intercrop, the infestation is negligible.

Q. How can ESB be chemically controlled?

ANS. At the time of planting, use Lesenta @ 150 gm/acre in 400 liters of water. Use this solution in the furrow as drenching for the seed cane. One insecticide can control two insect pests, that is, termites and ESB. Or at the time of first irrigation, use Coragen @ 150 ml/acre in 400 liters of water through drenching or use Chalorpyriphos/Phorate @ 12 kg/acre. Note: Coragen may not be used for ESB control because it is also used for top borer and one pesticide cannot be used two times on a single crop.

Q. How can top borer infestation be identified?

ANS. Top borer larvae are pale white. After wandering, they enter the midrib, leaving red marking, usually in the second to fifth leaf from the top. The caterpillar reaches the central core of the spindle through unfurled leaves. Its tracks are visible in the form of shot holes on the unfurling of leaves. The larva feeds by boring into the narrow central core towards the growing point and nibbling on the inner half of the leaf surrounding the feeding zone. As a result of biting across the spindle, a number of shot holes are formed in the leaf. Larvae enter the cane and damage the growing point. They feed on internal tissue; as a result, side shoots develop and give rise to a bunched top. Dead heart, when formed, is reddish brown, appears charred and cannot be easily pulled out. Activity of the top shoot borer starts with the onset of the monsoon rains.

Fig. 22

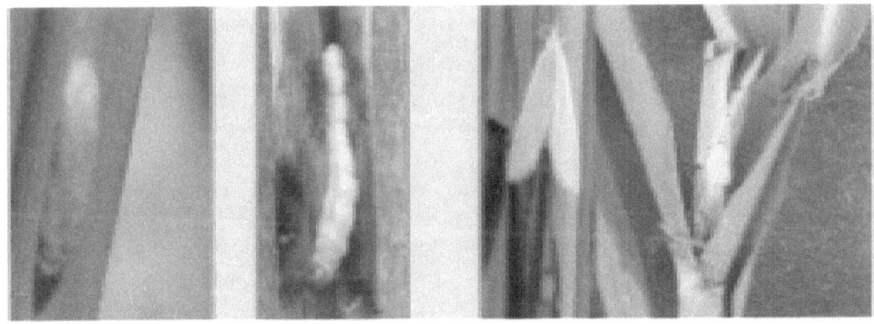

Q. How many generations affect the sugarcane crop?

ANS. The pest has five generations in the sub-tropical region:

I Brood	—	March second week to May third week
II Brood	—	May second week to June fourth week
III Brood	—	June first week to August first week
IV Brood	—	August first week to September third week
V Brood	—	September third week to February first week (over wintering of larvae)

Q. What is the extent of top borer damage noticed in sugarcane?

ANS. Heavy losses have been noticed due to top borer infestation:

Yield loss of 18–44%

Sugar loss of 0.1–2.10%

Q. How can top borers be controlled?

ANS. In the subtropical region, in the last week of May, use Coragen @ 150 ml per acre in 400 liters of water and drench near the root zone of the plant in a dry field and irrigate within 24 hours of drenching the insecticide. If drenching is not possible, then use Carbofuran 3G @ 30 kg/ha at June-end or before July 5, but the fertilizer must not be mixed with insecticide even in granular form. This decreases the effect of the insecticide on the crop and is also very harmful to the health of the laborer/grower.

Q. Can a foliar spray of urea and insecticide reduce infestation of insect pests?

ANS. On plant crop 90 DAP, use foliar spray of urea @ 5 kg/acre or NPK @ 1 kg/acre. This may be mixed with insecticides, Decis @100 ml/acre or Roket @ 500 ml/acre in 400 liters of water. It definitely controls insect pests such as ESB, white flies, mites and foliage cutter insects. It also speeds up the growth of the crop because it instantly fulfills nutrient requirements.

Q. Is bio control of top borer possible?

ANS. Maize crop can be grown along the border of the sugarcane field. The adult top borer is attracted to the maize and lays eggs on it. These lines must then be harvested and used as fodder for animals and from May, at fortnightly intervals, *trichogramma japponicum* parasite may be applied @ 50,000 eggs/ha for effective low-cost control of top borer.

Q. What is false attack of top borer?

ANS. When nitrogen is used at the time of top borer infestation, the crop achieves fast growth. The larvae try to reach the apical point, which grows upside down. This type of attack is called false attack. In such type of infestation, the apical portion of the plant seems like it has been burnt but after an interval, normal growth reappears.

Q. What is "Bunchy Top" in sugarcane?

ANS. Due to top borer attack, the apical point of sugarcane is damaged and the plant absorbs nutrients, so to grow the plant has to sprout the next eye buds and grow on top in a bunch. This is called Bunchy Top.

Fig. 23

Q. How can top borer infestation be identified from outside the field?

ANS. The infested plant leaf midrib has an identical symptom of top borer attack. After hatching, the larvae enters the midrib and moves downward. After a few days, this path turns red. It can be confirmed that fields that show this type of symptoms have been attacked by top borers.

Q. What is internode borer?

ANS. The caterpillars have a white body with dark spots and a brown head. They bore at the nodal region and enter the stem. Their feeding makes the tissues red. The entrance hole is usually plugged with excreta. A single larva can attack a number of nodes.

Q. What is stalk borer?

ANS. Soon after hatching, the larvae of stalk borer (*chilo auricilius*) wander for some time on the surface of the leaf and move downward to the central whorl where they feed by scraping the leaf sheath. Longitudinal orange yellow streaks are observed from tip to base on both sides of the midrib. The third instar larvae bore into shoots and internodes of the canes and feed on soft tissues.

Q. How can loss by stalk borers be reduced?

ANS. To reduce the infestation of stalk borers, the following steps may be followed:

1. De-trash the standing crop.
2. Harvest the water shoot.

3. Properly use water drainage system.
4. Use parasite cotesia flavipes @ 800–1,000/acre from July to November on a weekly basis.

Q. What is Gurdaspur borer?

ANS. This borer (*acigona steniellus*) was first noticed in Gurdaspur (Punjab). The larvae of this pest always enters the third/fourth internode from the apical part and moves upwards cutting the xylem of the stem. Due to this, it cannot be controlled with insecticide.

Q. When does Gurdaspur borer infestation appear?

ANS. The infestation of Gurdaspur borer appears from July-end to September-end or occasionally the first week of October.

Q. How can Gurdaspur borer infestation be identified from outside the field?

ANS. The infestation of Gurdaspur borer is found in bunches and it multiplies in double ratio. On infested clumps, all cane stems are healthy but the top turns dry. A strong wind or contact by an animal can break the top. On the infested plant de-trashing shows internode cut spots in the shape of a spring indicating the movement of larvae, which always attack on upper third or fourth internode.

Q. How many stages of Gurdaspur borer can be found on cane and which is more dangerous?

ANS. Two stages have been noticed on cane:

1. Gregarious stage and
2. Solitarious stage.

The gregarious stage is most dangerous to the cane since 10–20 larvae can enter a single cane in this stage.

Q. How can Gurdaspur borer be controlled?

ANS. This insect pest cannot be controlled using chemicals because its larvae cut the plant xylem and insecticides cannot reach the larvae in the cane. Thus, only a mechanical method is effective. The first four internodes of the affected plant should be cut and destroyed or used in fodder but these should not be thrown in the field or near it.

Q. What is *pyrilla*?

ANS. This is commonly known as leaf hopper (*pyrilla perpusilla*), these nymphs are pale brown with a pair of wax-covered anal processes. High humidity in May–June, heavy manuring and irrigation favor multiplication of the pest. Nymphs and adults suck sap from the leaves. In severe cases, the leaves fade and dry up. Plants present a sickly and blighted appearance. Insects excrete honeydew on which sooty mold develops. Since sucrose content is reduced, the quality of jaggery is drastically affected.

Q. How can *pyrilla* be controlled using the bio-control method?

ANS. Before infesting sugarcane, *pyrilla* usually attack fodder crop (*jawar*). When the parasite of pyrilla has developed in fodder crop, the best and cheapest method is to take the leaves containing the parasite (*epiricania melanoleuca*) of the fodder crop and spread it in the infested cane crop. If the parasite have not developed in nearby crop, then release another parasite which can be developed in a laboratory—*tetrastichus* or *chilocorus pyrillae*.

Q. Can the mechanical method be used to control *pyrilla*?

ANS. Yes. Grow a few lines at a short distance in the field and irrigate frequently using nitrogenus fertilizer on these lines. All egg laying should be along these lines and the leaves containing the eggs should be plucked and destroyed.

Q. What is mealy bug?

ANS. Its scientific name is *saccharicoccus sachhari*. The adult female is small, pinkish and has an oval well-segmented body with white waxy coating. Young ones are often produced parthenogenetically. Eggs hatch within a few hours and parthenogenetic reproduction helps rapid multiplication. Colonies of mealy bugs are found at the lower nodes of young cane and remain protected by the leaf sheath. Nymphs and adults suck sap and reduce the vitality of the crop, excrete honeydew on which sooty mould grows, due to which internodes and leaves appear black and cane growth is retarded.

Q. How can mealy bugs be controlled?

ANS. Mealy bugs are mainly mango insect pests but in sugarcane their infestation has been observed as minor to severe. Its infestation starts from the corner or border of the field. Due to this, in the starting phase it can be easily controlled while in the later stage it cannot be controlled and the entire field may be damaged. In the starting, use neem oil @ 250 ml with immedachloprid @ 375 ml/ha or neem oil + accetamiprid @ 250 gm/acre in 500 liters of water or methyl parathion dust @ 30 kg/hect or phorate 10 gm @ 30 kg/ha in a wet field.

Q. What are scale insects?

ANS. Its scientific name is *melanaspis glomerata*. Freshly hatched crawlers are light yellow and lose their legs after settling on the internodes. Nymphs and females look like an encrustation, remain attached to the cane, keep sucking the sap and devitalize the cane. Varieties with persistent leaf sheath are more susceptible. Infested canes shrivel, internodes shorten, and juice content and quality reduce drastically. They are mostly found in waterlogged areas.

Q. What is white fly?

ANS. Its scientific name is *aleurolobus barodensis*. Both nymphs and adults suck sap from leaves which show characteristic yellow streaks in severe cases; the leaves dry and plants remain stunted. Cane juice becomes watery and the setting of jaggery is affected. Yield and sucrose content are drastically reduced. Crops raised in low-lying, waterlogged alkaline soils suffer more. Peak activity is noticed in the post-monsoon period.

Q. What is black bug?

ANS. Its scientific name is *blisus gibbous*. Both nymphs and adults suck the sap from the leaves and mostly stay mid-sheath. The nymphs are brown or red while the adults turn red and have white wings. This insect pest mainly attacks ratoon crops from April to June; sometimes, its attack may be observed on plant crops from September to October.

Q. How can black bugs be controlled?

ANS. In April, May and June, foliar spray of insecticide must be used. Select any one from Decis 100 @ 250 ml/ha or imidacloprid @ 375 ml/ha or Roket @ 1.25 liter/ha in 800 liters of water and mix a pouch (10 ml) of shampoo for breaking the surface tension of water. The solution must be equally spread on the leaf after spraying. The foliar spray gives much better results when used with urea and insecticide in a wet field or just irrigate the field after the foliar spray.

Q. What is army worm?

ANS. Leaf-feeding insects are found on sugarcane throughout the world. The most important pests in this group are the caterpillars of various moths (lepidopterans). The caterpillars of the species of *spodoptera* and *mythimna* are associated with crop damage in Africa, Asia, Australasia, Malaysia and South America. The larvae hatch from eggs laid on the host or in surrounding grassed areas and move to young cane crops where they feed on the leaves. Complete defoliation can occur but the crop may recover if growing conditions are favorable. Repeated defoliation can lead to a serious loss in yield.

Q. What are some army worm control measures?

ANS. Chemical control is normally not advised as, most often, the damage has occurred before the problem is recognized.

Q. What are the insect pest-wise recommendations for insecticide and their dosage?

ANS. Details are in the table as under:

Table 14

S. No.	Name of insect pest	Time of infestation	Control measures (Chemical)
1	ESB (*chilo infuscatelus*)	March–June	Lesenta @ 150 gm/acre at the time of planting or Decis 100ml or imidacloprid 125 ml/acre should be used 60 and 75 DaP with urea or chlorpyriphos/phorate @ 12 kg/acre after first irrigation.
2	Root borer (*Emmalocera depressella*)	June–September	Quinalphos @ 2 liter/acre minimum twice.
3	Stem borer (*chilo auricilus*)	September–January	Quinalphos @ 2 liters/acre to be mixed in 30 kg sand and used in root zone in most moist parts of field.

Continued

4	Gurdaspur borer (*besetia stenielus*)	July–October	Only mechanical method applicable.
5	Top borer (*tryporyza nivella*)	July–September	Coragen @ 150 ml/acre in 400 liters of water at May-end or carbofuran 3G @ 12 kg/acre at June-end or in the first week of July.
6	*Pyrilla perpusilla*	July–September	Dusting of Methyl parathion @ 12 kg/acre.
7	Black bug (*blisus gibbus*)	April–June	Foliar spray of Decis 100 @ 100 ml/acre or imidacloprid @ 125 ml/acre or Roket @ 500 ml/acre in 250 liters of water.
8	Termite (*odontotermes obesus*) and White grub	February–October	Lesenta @ 150 gm/acre or Dontatsu @ 150 gm/acre or imidacloprid 125 ml/acre at the time of planting through drenching in furrow or seed treatment.
9	Woolly aphid	April–July	Foliar application of Decis 100 ml or imidacloprid 125 ml or accetameprid 250 gm + Neem oil 250 ml/ acre.
10	Mealy Bug (*saccharicocus sachhari*)	April–July	Spray imidacloprid/acetamiprid + neem oil on affected clumps and use Phorate @ 10 kg/acre in the whole field and irrigate immediately.

Q. How many insect pests in sugarcane can be controlled by the bio-control method?

ANS. In sugarcane crops, all borers, termites, black bugs, woolly aphids and *pyrilla* may be controlled through the bio-control method.

Q. How can predators/parasites be applied in the crop?

ANS. Insect pest-wise month-wise effective control is as under:

Table 15

S.No.	Name of insect pest	Time of infestation	Control measures (Bio)
1	ESB (*chilo infuscatelus*)	March–June	*Trichogramma chilonis* @ 50,000 eggs/acre per week.
2	Root borer (*Emmalocera depressella*)	June–September	*Trichogramma chilonis* @ 50,000 eggs/week.
3	Stem borer (*chilo auricilus*)	September–January	*Cotesia flavipes* @ 800–1000/acre on weekly basis from July to November.
4	Gurdaspur borer (*besetia stenielus*)	July–October	*Trichogramma chilonis* @ 50,000 eggs/acre per week from July to September.
5	Top borer (*tryporyza nivella*)	July–September	*Trichogramma japponicum* @ 50,000 eggs/acre per week and in the month of February use light trap @ 32 trap/ha.
6	*Pyrilla perpusilla*	July–September	*Epricania melanoleuca* or *Metarizium enasoply* or *tetrasticus* or *chilocorus pyrillae*
7	Black bug (*blisus gibbus*)	April–June	Spray of *beuvaria bassiana*.
8	Termite (*odontotermes obesus*)	February–October	*Beuveria bassiana* and *metarizium*. Prepare a culture in cow dung and use during land preparation and in July.
9	Woolly aphid	April–July	Dipha parasite may be developed in their colonies.

Q. What is IPM?

ANS. Integrated Pest Management (IPM) is an ecosystem-based strategy that focuses on long-term prevention of pests or their damage through a combination of techniques such as biological control, habitat manipulation, modification of cultural practices and use of resistant varieties. Pesticides are used only after monitoring indicates they are needed and according to established guidelines. Such treatments aim to remove only the target organism. Pest control materials are selected and applied in a manner that minimizes risks to human health, beneficial and non-targeted organisms and the environment.

Q. How can nematodes damage the sugarcane?

ANS. Plant-feeding nematodes are ubiquitous in farmlands and yet most farmers are unaware of their presence. Most are very small and worm-like (less than 2 mm in length) and not visible to the naked eye. Life stages include an egg, four juvenile stages and the adult stage. The plant-feeding nematodes associated with sugarcane are represented by more than 300 species worldwide, the most important of which are the root-knot nematodes, *meloidogyne javanica* and *meloidogyne incognita*.

Q. What are the symptoms of nematode infestation?

ANS. Nematodes feed on the contents of the cells of the roots and, where large populations occur, cause extensive damage and limit the functioning of the root system. Symptoms include reddish purple or purplish-black lesions on the roots. The roots become stubby, coarse and brittle; the tips may be enlarged, as discreet galls or as elongated swellings. The root system is sparse and appears dark.

Q. How can nematodes be controlled?

ANS. The most effective control option for reducing the impact of nematodes on sugarcane, but one that poses the greatest hazard to the environment, is the use of a nematicide. For effective control of nematodes Aldicarb GR or carbofuran 3G @ 30 kg/ha can be used.

BIO FERTILIZER

Using bio fertilizer is an economic way to increase crop yield and prevent degradation of soil health, which is why details of such practices become relevant for readers.

Q. What are bio-fertilizers?

ANS. Fertilizers prepared by living organisms are called bio-fertilizers.

Q. What is the classification of bio-fertilizers?

ANS. Bio-fertilizers have been categorized as under:

1. Organic matter decomposer — *Trichoderma virdie, cellulomonas arthrobacter*
2. Nitrogen-fixing bacteria — *Rhizobium azospirillum*
3. Phosphorus solubilizing — *Mycorrhiza, Bacillus pseudomonas*
4. Potash mobilizing — *Frateuriaurentia*

Q. What is the name of the nitrogenous bio-fertilizer?

ANS. Nitrogen is an essential nutrient for plant growth. About 80,000 ton of nitrogen is available in the environment but in its naturally occurring form, plants cannot access it. So, atmospheric nitrogen is changed by *azotobacter* bacteria into an accessible form in the node of the legume plant's root. These bacteria convert nitrogen into nitrate and store it in the root of the legume plants. If the legume crop is used as green manure, then the quantity of inorganic fertilizers can be reduced and crop yield increased by 5–10%.

Q. How can bio-fertilizers be used?

ANS. It can be used in crops at different times and through different methods as under:

1. Seed treatment: The recommended dose of bio-fertilizer mix in well-decomposed cow dung/FYM (200–300 kg/acre) should be used on the evening before planting and the next day before sharp

sunlight, the cane can be planted. Alternatively, 5 kg Azotobacter + 5 kg PSB both dissolved in 150 liters of water with 5 kg jaggery can be used to treat each hectare of the cane seed.

2. Seed treatment: The recommended dose of bio-fertilizer can be mixed in jaggery/sullary and used to cover seed grains.

3. After planting: The recommended dose of bio-fertilizer must be properly mixed in well-decomposed cow dung (200–300 kg/acre) and used in lines before first inter-culturing. After using the mixture, weeding/inter-culturing must be done immediately. This method should be used in the evenings on a wet field.

Q. What is mycorrhiza?

ANS. Mycorrhizas are symbiotic relationships between fungi and plant roots (the term literally means "fungus root"). Perhaps more than 80% of the species of higher plants have these relationships, and in several cases they have been shown to be important or even essential for plant performance. As the American plant pathologist, Stephen Wilhelm, said: "*in agricultural field conditions, plants do not, strictly speaking, have roots, they have mycorrhizas.*" (*The Microbial World: Mycorrhizas*, Jim Deacon)

Q. Is mycorrhiza available in the market?

ANS. Mycorrhiza fungi are available in the market with a different trade name, that is, Mycorrhiza.

Q. What is the mycorrhiza dose for sugarcane?

ANS. To increase the yield of sugarcane crops, mycorrhiza @ 5 kg/acre should be used at the time of planting or after the first weeding and this should be followed with earthing up.

Q. What is the benefit of using bio-fertilizers?

ANS. The benefits of using bio-fertilizers are as under:

1. It increases germination percentage.
2. It reduces the quantity of inorganic fertilizers required by about 25%.
3. It increases the quantity of micro-organisms in soil.

4. It improves the quality of the final product.
5. It makes nutrients easily available to plants.
6. It improves soil texture and fertility.
7. It helps prevent diseases.

INTERCROPPING AND CROP ROTATION

This practice is beneficial to sugarcane growers since it increases their returns and also helps upgrade soil health when the grower selects legume crops for intercropping or green manure. Crop rotation helps increase the yield and maximizes the use of soil nutrients without any extra cost.

Q. What is intercropping pattern?

ANS. When crops do not compete with each other and grow equally at the same time in the same field, it is called intercropping. Intercropping with sugarcane is a profitable practice. It can be done both with leguminous and non-leguminous crops. Intercropping has the additional advantage of inter-culture and earthing up.

Q. Which crop has no impact on sugarcane?

ANS. If the following are grown as intercrops, they reduce the infestation of insect pests as well as diseases like wilt and red rot (secondary infestation): 1. sugarcane + potato (autumn); 2. sugarcane + wheat (autumn); 3. sugarcane + mustard (*toria*) (autumn); 4. sugarcane + lentil (autumn) 5. sugarcane + green gram (spring); 6. sugarcane + black gram (*urad*) (spring); 7. sugarcane + soya bean and 8. sugarcane + coriander/garlic/onion (autumn).

Q. Can red rot disease be reduced/controlled using a particular crop pattern?

ANS. Yes, if rice is grown before planting sugarcane, it controls the infestation of red rot in sugarcane.

Q. What is the best method for intercropping onion and sugarcane?

ANS. The trench method of planting is the best method. In this method, two lines are created 120 cm apart and this space can be used for four lines of onion.

Q. While intercropping, what should the line-to-line and plant-to-plant distance of the onion crop be?

ANS. Line-to-line distance should be 15–20 cm and plant-to-plant distance should be 10–15 cm.

Q. What is the balanced dose of fertilizers for onion crop?

ANS.

1. Nitrogen — 120 kg/ha
2. Phosphorus — 60–70 kg/ha
3. MOP — 60–70 kg/ha
4. Sulphur — 20–30 kg/ha

Q. What are some onion varieties that have a good yield?

ANS. Red color—Pusa red, Pusa ratnar and arka Kalyan.

Yellow color—Early grano and Early yellow globe.

White color—Patna white, UD-102 and white globe.

Q. Why are intercropping and crop rotation important for sugarcane?

ANS. Due to intercropping, sugarcane gets more inter-culturing practice and soil has pulverization for good development of root system. Some crops help control insect pest infestation or secondary disease. Crop rotation is also very beneficial for increasing yield and preventing degradation in variety.

Q. What is the benefit of crop rotation?

ANS. Growing the same crops in the same site year after year reduces soil fertility and can encourage a buildup of pests, diseases and weeds in the soil. Crops should be moved to a different area of land each year. Crop rotation also helps a variety of natural predators to survive on the farm by providing diverse habitats and sources of food for them.

Q. What are some patterns of crop rotation?

ANS. The following crop rotation may be used for better results:

1. For two years 1. maize–potato–sugarcane–ratoon; 2. maize–sugarcane–ratoon–wheat; 3. paddy–sugarcane–ratoon–wheat and 4. cotton–sugarcane–ratoon–rabi jowar.

2. For three years 1. *lowa* (fodder)–potato–sugarcane–wheat; 2. paddy–gram–sugarcane–ratoon–wheat; 3. cotton–sugarcane–ratoon; 4. paddy–*toria*–sugarcane–ratoon–wheat; 5. maize–wheat–sugarcane–ratoon–wheat; 6. paddy–sugarcane–ratoon–wheat; 7. cotton–sugarcane–ratoon–wheat; 8. cotton–sugarcane–ratoon–gram and 9. paddy–groundnut–jowar–ragi–sugarcane.

Q. How does crop geometry help?

ANS. Adoption of the crop geometry concept for harvesting ensures more solar energy efficiency and helps plants convert it effectively into food energy with the help of the crop.

RATOON MANAGEMENT

All people who involved in sugarcane farming must know about all the practices related to ratoon for increasing yield.

Q. What is the first activity to be undertaken after harvesting the plant crop?

ANS. Trash burning or trash lining is the first activity to be undertaken after the plant crop is harvested. It can be completed within five days.

Q. What is the impact of stubble shaving on the ratoon crop?

ANS. Stubble shaving with a sharp *panga (balkati)* involves cutting the stubble at ground level to obtain uniform, vigorous and strong tiller growth. Mechanized cutting can also be done though ratoon management device (RMD).

Q. What is off-barring in ratoon crop?

ANS. Breaking the shoulder of side ridges with oxen plough or RMD to cut old roots and enhance the development of new roots for plant growth.

Q. Which fertilizer can be used for good growth?

ANS. After off-barring use DAP/NPK @ 100 kg/acre or Single Super Phosphate (Ramban) @ 300 kg/acre. If possible, mix the chemical fertilizer with well-decomposed organic manure @ 100 kg/acre and apply row to row. After the application of fertilizers, must be earthing up to prevent the fertilizer from evaporation or leaching in soil cultivator/tynner must be used to mix the fertilizer in the soil within the root zone, or use RMD. The activity should be completed within fifteen days of harvesting the crop.

Q. Is gap-filling for ratoon crop beneficial?

ANS. Yes, this activity is very beneficial for maintaining optimum plant population to get a higher yield. In this activity, excess tillers can be removed along with the root and soil portion and the gaps within rows filled. Alternatively, polybag nursery developed plantlets can be planted.

Q. How and when must urea be used to get the best result?

ANS. Urea @ 50 kg/acre must be applied near the root zone followed by earthing up by oxen plough/RMD within sixty days of harvesting.

Q. Why must foliar spray of urea be used in ratoon crop?

ANS. Foliar application of urea in wet conditions is very beneficial for yield increase. To control sap-sucking insect pests such as black bug, white fly, *pyrilla* and leaf cutters, mix the insecticide in the solution and use.

Q. How can ratoon crop yield be increased?

ANS. After first weeding in ratoon crop, at the time of fertilizer application if Mycorrhiza @ 5 kg/acre or a mixture of humic acid and amino acid @ 10–12 kg/acre is also added and followed with earthing up, this helps increase yield. The field must be irrigated or must have sufficient moisture. Applying humic acid mixture with NPK/DAP or urea increases tillering and fast growth. The mortality of tillers also decreases. Mycorrhiza/humic acid and amino acid act as catalysts converting unavailable forms of nutrients into available forms.

Q. Can mycorrhiza or humic acid mixture be used in plant crops?

Ans. It can be used in plant and ratoon crops. For best results, the humic acid mixture must be properly mixed with fertilizer after first weeding and followed by earthing up. In plant crop, it can be used at the time of planting, with NPK/DAP fertilizer. In plant crop, after applying humic acid and amino acid combination it has been observed that the root system develops and absorbs the nutrients as per plant requirements.

Q. What is the water requirement at different crop stages?

ANS. The sugarcane crop has four stages. Water requirement in each stage is as under

1. Germination phase: 300–320 mm (sub-tropic); 200–250 mm (tropical region).
2. Tillering phase: 550–600 mm (sub-tropic); 300–350 mm (tropical region).

3. Grand growth phase: 1,000–1,100 mm (sub-tropic); 500–550 mm (tropical region).
4. Maturity phase: 650–700 mm (sub-tropic); 300–320 mm (tropical region).

DRIP IRRIGATION IN SUGARCANE

In India, especially in the north, flood irrigation is used for crops. The crop absorbs its requirement of water and the remaining water is either leached or evaporated. Such loss can be prevented with drip irrigation, which also boosts yield.

Q. What is drip irrigation?

ANS. Drip irrigation is a technology with great potential for improving the efficiency of water use and for increasing crop production and food security by enabling agriculture on marginal land.

Q. What impact does drip irrigation have on sugarcane?

ANS. Sugarcane yields have been increased by 15–25% using drip irrigation instead of flood irrigation.

Q. How much water can be saved by this method?

ANS. Using drip-irrigation system in sugarcane, saves water, energy and labor as under:

1. Water   40%
2. Electricity   28%
3. Labor   87%

Q. What is the usage pattern of fertilizers in the drip-irrigation system?

ANS. With this method, the usage of fertilizers should be as follows:

1. Nitrogen   95%
2. Phosphorus   45%
3. Potassium   80%

Whereas in flood irrigation, this ratio is as under:
1. Nitrogen   30–50%
2. Phosphorus   20–25%
3. Potassium   50–55%

Q. How can sucrose content be increased through drip irrigation in sugarcane?

ANS. Since fertilizer is used in the root zone through water in soluble form, the root easily absorbs it and plants' nutrients requirements are fulfilled without any scarcity. Phosphorus and potassium play a vital role in sucrose accumulation because potassium is a quality nutrient.

Q. What is the difference between drip and conventional irrigation?

ANS. A comparison of the two are as under:

Table 16

S. No	Parameters	Expenses/Acre (Rs) Conventional	Expenses/Acre (Rs) Drip System	Remarks
1	Land preparation			
	MB Plough	3,000	3,000	
	Disc Plough	2,600	2,600	
	Harrowing	1,800	1,800	
	Tillering	1,200	1,200	
	Furrowing	1,000	1,000	
2	Seed cane cost	10,500	10,500	@ Rs 300/ Quintal
	Transportation of seed	525	525	@ Rs 15/ Quintal
	Loading/Offloading	200	200	
	Planting cost/labor	1,800	1,800	
3	Fertilizers			
	DAP/NPK	2,600	2,600	@ Rs 1,300/ bag

Continued

	Soluble fertilizer 19:19:19 @ 4 kg/acre		1,500	@ Rs 60/kg
	Urea	1,600	0	@ Rs 800/bag
	Zinc/Sulphur	350	350	@ Rs 35/kg
	Other			
4	Weeding			
	Manual	6,800	3,400	@ Rs 1,700/acre/weeding
	Chemical	2,000	0	
5	Foliar nutrient application	300	0	
	Labor	200	0	
6	Insect pest management			
	Lesenta (Termite/White grub)	2,000	2,000	
	Lesenta (ESB)	0	0	
	Coragen (Top Borer)	2,000	2,000	
	Dust (Grass hopper/Pyrilla)	200	200	
7	Irrigation	3000	600	
8	Cane Propping	1,500	1,500	
9	Harvesting & Loading	7,500	9,375	@ Rs 25/Quintal
	Transportation	4,500	5,625	@ Rs 15/Quintal
	Total	57,175	51,775	
10	Drip system cost	0	35,000	One-time cost
11	Production (Quintals)	300	375	
12	Income @ Rs 280/Quintal	84,000	1,05,000	
13	**Net Income**	**26,825**	**18,225**	

ADVANTAGE OF NEEM IN SUGARCANE

Since the past decade, use of chemicals against insect pests has increased. The residual effect of insecticide is also a challenge in fodder. The use of neem in soil helps improve soil health and control soil pests and nematode naturally.

Q. What is the scientific name of neem?

ANS. Scientific names: *azadirachta indica A. Juss.* Family: meliaceae

Common names: *neem, margosa, nim, nimba, nimbatiktam, arishtha, praneem*

Q. What is neem cake?

ANS. Neem cake is a strange name for the pulp that is left after extracting neem seed oil from kernels. It is used as fodder sometimes. However, the most common and recommended use is as a soil amendment and fertilizer for preventive measures against termite and other soil insects and nematodes.

Q. Why does the use of neem cake prevent damage from nematode and termite?

ANS. Neem cake has poisonous ingredients such as *nimbin, nimbolide, nimbinin* and *selenin* due to presence of which nematodes and termites cannot damage crops.

Q. Should neem cake be mixed in urea?

ANS. Neem cake should be mixed in urea as it is very beneficial to the crop. Neem cake absorbs nitrogen from urea and makes it available to crops for a long time, whereas urea used alone could evaporate or leach in soil with water.

Q. How can termites be controlled with neem?

ANS. In a 20 kg earthenware pot, mix about 5 kg neem leaves and about 2 kg *madaar (aakh)* leaves with 10 liters of cow urine and keep the lid of the pot closed for fifteen days. This mixture is then ready for use. After filtering the solution, dilute it in 150 liters of water for one acre as the basal drench on seed cane in the furrow during planting or as foliar application on vegetable/flowers/or any other crop to protect from leaf-chewing or sap-sucking insect pests.

Q. How can this mixture be used in sugarcane?

ANS. The ready-to-use mixture can be diluted 1:10 and filtered with a wire mesh or cloth for foliar spray application on sugarcane or other crops.

Q. How does neem improve yield and soil structure?

ANS. All parts of the neem plant are very beneficial. It can be used as mulch, as a compost ingredient or as a soil amendment. Neem can be used to reclaim marginal soils. It can bring acidic soils back to neutral; the deep tap root can break through hard layers, mine the subsoil for nutrients and bring them to the surface. Growing neem trees improves the water-retention capacity and nutrient level of soils. Again, this is a very promising use of the neem plant. It could make a huge difference, not only in third-world countries, but also on abused agricultural soils. When neem cake is used in soil, it controls soil insects and harmful nematodes and improves the soil structure.

Q. What is the chemical composition of neem leaves?

ANS. Fresh neem leaves have the following ingredients:

Protein	7.1%
Fat	1.0%
Fiber	6.2%
Carbohydrate	22.9%
Calcium	510 mh/100 gm
Phosphorus	80 mg/100 gm
Iron	17 mg/100 gm
Thiamine	.04 mg/100 gm
Vitamin C	218 mg/100 gm
Minerals	3–4%

CANE MARKETING IN UTTAR PRADESH

When we discuss the sugarcane crop, then knowledge of procurement must also be updated. This is why we have devoted a chapter to cane development in this manuscript.

Q. How is the requirement of a sugar mill be calculated?

ANS. The cane requirement for the sugar mills in Uttar Pradesh (UP) is decided under Section-12 of the UP Sugarcane (Regulation of Supply and Purchase) Act-1953. Keeping in view their geographical condition/crushing capacity and their allotted areas, all the state's sugar mills have to increase the drawl by a minimum 60–70% over the previous year.

Basic Quota

Q. How is basic quota calculated?

ANS. Basic quota is to be calculated using the following formula (UP Cane Bonding Policy):

A. The certified details of each cane grower must be maintained based on the revenue records of a cane society and every change in it should be amended and updated.

B. Only the member of a co-operative cane development society/co-operative sugar mill society, who has land according to land revenue records, will get bonding facilities for their sugarcane. Lessees of railway, revenue, and forest & irrigation departments would also get bonding facilities based on the previous year's supply after physical verification of their sugarcane area.

C. The basic quota of a cane grower is to be calculated on the average supply of two crushing seasons. For growers who enrolled in the previous crushing season and supplied sugarcane for a year, their one-year sugarcane supply will be their basic quota.

D. If the requirement of a sugar factory is not met by the above method, then the balance is to be met by general enhancement after obtaining the recommendation of the concerned deputy cane commissioner.

E. The basic quota is to be calculated as up to 85% of total sugarcane produced but this total quantity should not be more than the total requirement of the sugar mill. If the bonding is more than the requirement it is to be reduced on pro-rata basis and brought down to the requirement of the sugar mill.

Q. What happens if farmers have excess cane from their basic quota?

ANS. In any sugar mill area, if the cane grower's total basic quota is insufficient to meet the requirement of the sugar mill, then the difference is met by additional bonding. For additional bonding, general category farmers are charged Rs 2 per quintal, small and marginal farmers are charged Rs 1 per quintal and SC/ST category farmers are charged 50 paisa as security money.

Q. When is this security money for additional bonding returned to growers?

ANS. On the supply of bonded sugarcane by these farmers, 50 paisa per quintal is held back as administrative fee and the rest of the security amount is paid to them.

Q. What are the bonding limits for different types of farmers?

ANS. The bonding limits of cane growers are based on the classification of their land-holdings and the quantity is fixed as follows:

Marginal farmers — 1 hectare (maximum 750 quintals or/and in case of yield enhancement, maximum 1,000 quintals).

Small farmers — 2 hectares (maximum 1,500 quintals or/and in case of yield enhancement, maximum 2,000 quintals).

General farmer — 5 hectares (maximum 3,750 quintals or/and in case of yield enhancement, maximum 5,000 quintals).

Maximum quality is based on cane area (hectare) × 750 quintal and in case of yield enhancement, maximum 5,000 quintals or whichever is less shall be taken.

(Universities, Ganna Beej Nigam, Sugar Mills, Agriculture Department of Central or State Government, Jails, registered co-operative societies having land and/or agricultural farms are exempted from the upper limit of bonding although prior permission of the cane commissioner is required.)

Q. Why is the bonding list displayed in each village?

ANS. A grower-wise bonding list has to be displayed at a public place between August 1 and 15 by the concerned *gram pradhan*/sugarcane supervisor and mill employee under the supervision of the cane development inspector and cane manager of the concerned sugar mill. the dates of display must be advertised properly and objections to the bonding list from cane growers are invited and recorded in a register. the senior cane development inspector and cane manager are authorized to check and correct the list on the basis of objections filed by the growers

Role of Calendar

Q. What is the right time for pre-calendar and final calendar distribution?

ANS. Pre-calendar distribution in the western region is to be completed by October 15 and in the central region by October 31. Objections received from the growers (except surveys) should be disposed of chronologically. The cane society must enter into an agreement in the prescribed pro-forma with the cane grower for the determined quantity of bond. Final calendars are to be prepared in triplicate, with one copy given to the grower, another to the society and the third to the sugar mill.

Q. What happens if a famer has land in different villages?

ANS. The details of farmers who are under the jurisdiction of one society but cultivate sugarcane at more than one place should be collected and entry bonding, by transfer, should be operated from one place only. The concerned block in-charge will check the transfer entry on the basis of land revenue records and after the approval of the senior cane development inspector, another cane development inspector will accept the transfer entry.

Role of Indent

Q. What is the procedure of indent?

ANS. The indent of cane supply must be sent to the society by the sugar mill at least four days earlier. The indent quantity is to be fixed on the basis of balance indent and daily crushing capacity. The mill and the secretary are responsible for ensuring that every out-center's daily weighment record is displayed on the notice board.

Q. What is equitable cane purchase?

ANS. Equitable cane purchase should be managed by supply head/cane head of sugar factory at every purchase center. The District Cane Officer (DCO) must also monitor this in fortnightly Cane Implementing Committee (CIC) meetings.

Q. What is the procedure of burnt cane supply?

ANS. After an FIR has been filed in the police station, grower must submit a copy of the same to society/mill for issuing burnt cane supply tickets. The detail of such slips must be maintained separately with the joint signature of the secretary of the society and the Cane Head.

Q. How can ratoon and early cane be supplied?

ANS. Ratoon and autumnal planting cane must be supplied on the basis of availability up to a maximum of January 31 or before this date up to the time when the crushing is possible only on the supply of ratoon and autumn. The supply of ratoon sugarcane shall be made up to January 31 and to the 60% limit of total bonding of sugarcane. The supply of extra available ratoon will be taken with the general variety sugarcane.

Q. What facilities are available to marginal growers in cane bonding policy?

ANS. Priority shall be given to marginal sugarcane farmers for supplying their plant and ratoon cane within 45 days of the commencement of the mill and from February 1 within forty-five days. Farmers having sugarcane bond of four bullock carts (60 quintals) shall be deemed marginal farmers.

Type of Reports

Q. What type of report is required on a daily basis?

ANS. Different types of reports are required on a daily basis during the crushing season to manage the out-center operation smoothly and to fulfill requirements. Updating the reports every morning and evening by block/zone/supply in-charge to the Head of Department (HOD) is important for the next day's planning and indenting to fulfil the requirement of factories and manage equitable purchase at mill gates and out-purchase center.

1. Daily morning reports.
2. Daily purchase reports.
3. Daily indent balance report.
4. Daily purchase and dispatch report.
5. Daily out-center balance report.
6. Daily out-center checking report.
7. Daily evening report.
8. Daily Hand-Held Computer (HHC) operation report.

AGRONOMICAL PRACTICES IN KENYA

Since I am currently in Kenya and aim to produce knowledge that is relevant for every reader of my book, I have devoted this chapter to the agronomical practices in Kenya.

Soil in Kenya normally lacks adequate amount of N and P and K are found only in some pockets in small amounts. Annual rainfall of up to 1,600 mm occurs over two rainy seasons (March to June and September to November) with minimum average temperatures of 14 °C and maximum average temperatures of 29 °C; the coolest months are July–August and the hottest months are December–January. Kenya's main cash crops are tea, coffee, maize and sugarcane. The sugarcane crop is harvested once every eighteen months because all crops are rain-fed; in some pockets, they are burned before harvest.

Q. What is the planting method used?

ANS. It varies from place to place but most frequently adopted methods are furrows of different kind, that is, end-to-end, overlapping, double rows and paired rows. The most common method is double row within single furrow.

Q. How can cane yield be increased?

ANS. Good land preparation means first ploughing with MB Plough then disc plough, harrowing if required and then one harrow repeated for good tilth. This helps in root development and improves the soil's properties.

Q. What is the seed rate per acre?

ANS. In Kenya, 3.0–4.0 ton/acre seed are used during planting.

Q. What is the row spacing?

ANS. Row-to-row space is generally 90 cm although it has now been increased to 120–150 cm because of new cane varieties with more tiller capacity. Plant population is maintained with wide row spacing.

Q. What is the dosage and time of fertilizer application?

ANS. Farmers usually do not use basal dose of fertilizer during planting. They use basal dose after first weeding @ 100 kg/acre DAP/NPK. This is due to continued rainfall and because more weeds could develop in basal use.

Q. What are the recommendations of the KALRO-SRI?

ANS. It recommends that phosphate fertilizers are used at the time of planting and nitrogenous fertilizers are for top dressing. The recommendations of all fertilizers are as under:

Table 17: N: 120 kg/ha; P: 80 kg/ha & K: 50 kg/ha

S.No.	Fertilizer	Plant	Ratoon
1	Phosphate		
	DAP (Bags/ha)	5 (250 kg)	5 (250 kg)
	SSP (Bags/ha)	9 (450 kg)	9 (450 kg)
2	MOP (Bags/ha)*	2 (100 kg)	-
3	Nitrogenous		
	Urea (Bags/ha)	4 (200 kg)	5 (250 kg)
	Calcium Ammonium Nitrate (Bags/ha)	7 (350 kg)	9 (450 kg)

*N.B.: MOP recommendation only for western zone

Source: KALRO-SRI monthly newsletter 2015

Q. What is the time for nitrogenous fertilizer application?

ANS. The first dose of urea may be used after second weeding and earthing up and the second dose of urea can be applied after third/fourth weeding up to eight months of crop age.

Q. How can herbicide be used in sugarcane crop?

ANS. The first dose of herbicide must be used within three days of planting and the second dose should be applied by the tenth month of crop age.

PART II

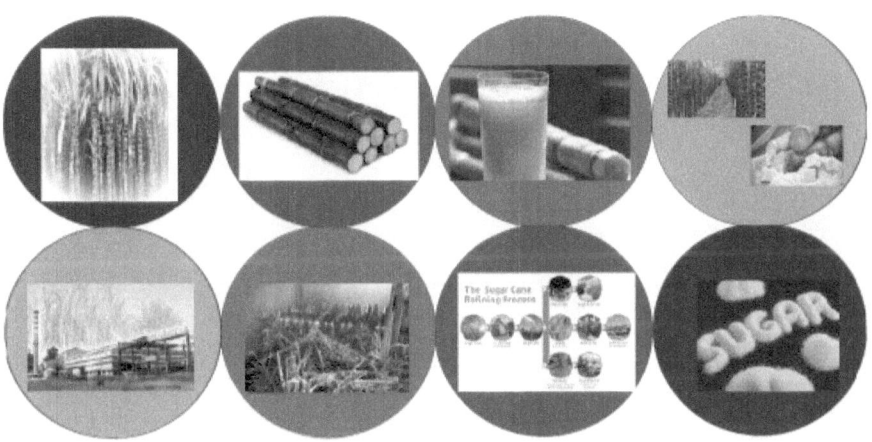

BASIC QUALITY CONTROL

This chapter useful to all professionals for general idea about some definitions related to sugar industry and general formula using in sugar industry for Milling control and Quality control.

Q. What is attenuation index?

ANS. This is part of the International Commission for Uniform Methods of Sugar Analysis (ICUMSA) sugar-rating process and refers to the light absorbance of a solution. It is tested at a specific wavelength and expressed in terms of that wavelength.

Q. What is brix-free cane water and how can we calculate it?

ANS. This refers to a measurement of the ratio of the mass of dissolved sugar to the mass of water in an aqueous solution.

Brix-free water is water that is present in cane and bagasse but is not available for dissolving sucrose in the cane. It is estimated that dry cane fiber has approximately 25% brix-free water. This water is generally removed by heating the fiber to evaporate the water.

Brix-free cane water or undetermined water = Cane - Fiber - Undiluted cane juice

Q. What is final bagasse?

ANS. Bagasse obtained after the first mill is called first mill bagasse. Depending on how many milling stages the sugar cane goes through, there may be second mill bagasse, third mill bagasse, etc. Post the diffusion stage, the bagasse is known as diffuser bagasse, and the very last type of bagasse, made after the dewatering stage is known as bagasse or final bagasse.

Q. What are insoluble solids?

ANS. The material present in the mixed sugar juice which does not dissolve and must be removed by sedimentation and/or filtration. In many cases, this material will collect at the bottom of sub-sides (clarifier) and be removed.

Q. What is intermixed cane?

ANS. This term refers to a method of carrying sugar cane, that is, to place many differing consignments on one carrier. Cane from various consignments mix together to make a blend of cane that possesses properties that are not representative of any of the consignments on board the carrier and should not be used for analytical purposes.

Q. What is invert sugar?

ANS. When sucrose is hydrolyzed, invert sugar is produced. This is a sugar mixture that is half glucose and half fructose.

Q. What is pol?

ANS. Sucrose content when expressed as a percentage. Most refined sugar has a very high pol, 99–100%. This measurement is called pol because it is determined by the polarization method.

Q. What is Java ratio?

ANS. Java ratio is the percentage ratio of the percentage of pol in cane to the percentage of pol in the first juice.

Q. What is boiling house & BH Recovery?

ANS. Boiling house is where juices are taken after sulphitation or phosphorization to be boiled down.

Boiling House (BH) Recovery refers to a percentage ratio describing how much pol is recovered in the form of sugar from the mixed juice.

Q. What is cane to sugar ratio?

ANS. It is a measurement that describes how many tons of cane are required to produce one ton of sugar.

Q. What is cold digester and crystal content?

ANS. Cold digester is a piece of equipment that disintegrates cane or bagasse in water to create a homogenous solution. It is used for analytical procedures.

Crystal Content is a measurement describing the mass of crystalline sugar present in a liquid.

Q. What is DAC extract and factor?

ANS. A liquid fraction that is decanted from the cane once it has been blended with water in the cold digester is DAC extract.

There are two types of DAC factors:

Brix Factor: This is the percentage ratio of the total brix in the mixed juice and the final bagasse to the total brix in the cane.

Pol Factor: The percentage ratio of total pol in the mixed juice and the final bagasse to the total pol in the cane.

Q. What is dextran?

ANS. It is a polysaccharides and series of polymers of glucose with α 1–6 linkages. It is a product of microbial infection of damaged cane cells. It is a gummy substance with molecular weight of 15,000 to 2,000,000. It is formed rapidly under conditions of acid pH, low brix and slightly elevated temperature.

It grows at an optimum temperature of 25–30 °C and generation time is twenty minutes and causative organism is *leuconostoc mesenteroides*.

Q. What is post -harvest deterioration in sugarcane?

ANS. The losses increase with the increase of duration of staling and it varies with varieties (Solomon).

Q. How can access the losses with staling period in sucrose and single cane weight?

ANS. The post-harvest losses evaluate in sucrose and single cane weight as under:

Effect of staling period on per cent juice sucrose

Variety	Staling period after harvest						
	0 Hrs	24 Hrs	48 Hrs	72 Hrs	96 Hrs	120 Hrs	Mean
CoC 671	17.63	17.32	17.01	16.43	16.05	15.53	16.66
Co 94008	17.03	16.84	16.56	15.95	15.46	15.16	16.17
Cose 92423	15.12	12.5	11.24	10.14	8.49	7.18	10.78

Effect of staling on single cane weight (kg)

Variety	Staling period after harvest						
	0 Hrs	24 Hrs	48 Hrs	72 Hrs	96 Hrs	120 Hrs	Mean
CoC 671	2.06	1.73	1.64	1.56	1.46	1.37	1.64
Co 94008	1.78	1.55	1.45	1.35	1.26	1.15	1.42
Cose 92423	1.59	1.38	1.27	1.18	1.1	0.97	1.25

Source: The BioScan (2014)

Q. What are the effects of dextran in the sugar manufacturing process?

ANS. The accumulation of dextran causes many process problems:

1. Low recovery,
2. Increase in viscosity,
3. Increase in molasses purity,
4. Filtration difficulties,
5. Crystal elongation.

Apart from this, the most significant problem is the effect of dextran on polarization readings because dextran has a very high specific rotation, $(\alpha)^{20}_d = 1,199$, which results in false sugar content and false juice purity.

Q. Which bacteria affect cane juice?

ANS. *Bacillus stereo thermophile* bacteria affects cane juice six–eight hours after harvesting the cane.

Q. What is the composition of sugarcane?

ANS. The composition of sugarcane is given below:

1. Water   69–75%
2. Sucrose   8–16%
3. Fiber   8–16%
4. Dissolved sugar   1–2%
5. Ash   0.3–0.8%
6. Organic matter   0.5–1.0%
7. Nitrogenous matter 0.02–0.05%
8. Others   0.65–1.02%

Q. What is the chemical composition of filter cake (press mud)?

ANS. The following nutrients may be available in the filter cake of sugarcane which is released by the sugar mill after producing the final product, sugar:

1. Nitrogen   1.50–1.69%
2. Phosphorus   2.0–2.29%
3. Potassium   0.4–0.44%
4. Sulphur   1.8–2.0%

5. Zinc   0.02–0.03%
6. Copper   0.01%
7. Manganese   0.02%
8. Magnesium   0.51%

Q. What is quality deterioration in sugarcane?

ANS. Once sugarcane is cut, it is subject to deterioration, largely due to the activity of microorganisms. This results in the loss of sugar and the formation of undesirable impurities. The extent of deterioration is determined by a number of factors, but in all cases the effect on the recovery of sugars and the processing of the cane is adverse. It, therefore, needs to be kept to a minimum.

Q. How many types of juices result from milling?

ANS. During this process, cane juice is converted into different types as under:

Absolute Juice: This juice only exists hypothetically and is the mass of the sugarcane minus the mass of fiber. It is not possible to ever completely extract all the sugar and liquid present in the cane.

Clarified Juice: Juice which has been clarified.

Diffuser Juice: Juice removed from the sugarcane or bagasse diffusers.

First Expressed Juice: The first juice extracted by the first two rollers.

First Mill Juice: The first juice extracted from the first mill.

Last Expressed Juice: The last juice extracted by the last rollers.

Last Mill Juice: The last juice extracted from the last mill.

Mixed Juice: The juice that is pumped to the juice scales from the extraction plant.

Press Water: The liquid removed when the diffuser bagasse is dewatered.

Primary Juice: The combined juices prior to treatment.

Residual Juice: The juice present in bagasse, apart from the juice in the first bagasse.

Secondary Juice: Diluted juice which is mixed with primary juice to make mixed juice.

Undiluted Juice: All the juice in the cane. Once again, this is fairly hypothetical as it is not possible to remove all juice in a useful fashion.

Q. What is magma?

ANS. The mixture of sugar crystals and warm sugar liquor created in the first stages of the refining process is magma.

Q. What is Massecuite?

ANS. Crystals and mother liquor removed from a vacuum pan as a liquid mixture is known as Massecuite. There are various grades of Massecuite, determined by their purity: A Massecuite, B Massecuite, C Massecuite, C-1 Massecuite, R1 Massecuite, R2 Massecuite, etc.

Q. What is milling loss?

ANS. A measurement describing the percentage ratio of pol (sucrose content) in bagasse compared to fiber in bagasse.

Q. What is molasses?

ANS. A thick, dark, sweet and highly viscous substance separated from sugar at the beginning of the refining process.

Q. What is mud?

ANS. This term refers to the sludge-like material that is cleaned from the lower regions of the sub-sides. This is made up of liquid and insoluble substances.

Q. What is non-pol and non-pol ratio?

ANS. Non-Pol: A term used to describe the brix minus any pol (sugar content.)

Non-Pol Ratio: This describes the amount of non-pol in sugar, the non-pol in final molasses and non-pol in mixed juice.

Q. What is normal mass?

ANS. The normal mass of sucrose should be 26,000 gm. This is determined by calculating the mass of dry sucrose, which when dissolved in 100 cubic centimeters of water at 20 °Celsius, and then read in a tub 200 mm long, reads 100 degrees on the saccharimeter scale.

Q. What is nutch sample?

ANS. A sample of molasses removed prior to the curing of the Massecuite.

Q. What are polysaccharides?

ANS. A complex carbohydrate's molecule where many saccharide molecules are bonded together.

Q. What is preparation index?

ANS. This is the percentage ratio of brix in ruptured cells, compared to the total amount of brix in the cane.

Q. What is purity?

ANS. The ratio of pol to brix. Essentially this term describes how much pure sucrose is present in a sugar sample.

Q. What is Super Saturation Coefficient?

ANS. Super Saturation Coefficient is the ratio which compares how much sucrose is present in a sample with the potential solubility of sucrose in the sample under constant conditions.

Q. What formulas are used in quality control?

ANS. During the milling and manufacturing process, many formulas are used as under:

CHEMICAL CONTROL FORMULA

Net Mixed Juice % Cane = Gross Mixed Juice % Cane - Dirt Correction

$$\text{Net Mixed Juice \% Cane} = \frac{\text{Net Weight of MJ} \times 100}{\text{Weight of cane}}$$

$$\text{Added water \% cane} = \frac{\text{Weight of Added Water} \times 100}{\text{Weight of Cane}}$$

$$\text{Bagasse \% Cane} = \frac{\text{Weight of Bagasse} \times 100}{\text{Weight of Cane}}$$

$$\text{Brix \% Cane} = \frac{\text{Tons Brix in Cane} \times 100}{\text{Tones Cane}}$$

$$\text{Brix \% Bagasse} = \frac{\text{Pol \% Bagasse} \times 100}{\text{Purity of LMJ}}$$

Fiber % bagasse = 100 - Brix % Bagasse - Moisture % Bagasse

$$\text{Tons Brix in MJ} = \frac{\text{Brix \% MJ} \times \text{Tons MJ (Net)}}{100}$$

$$\text{Tons Pol in MJ} = \frac{\text{Pol \% MJ} \times \text{Tons MJ (Net)}}{100}$$

$$\text{Tons Pol in Bagasse} = \frac{\text{Pol \% Bagasse.} \times \text{Tons Bagasse}}{100}$$

$$\text{Tons Brix in Bagasse} = \frac{\text{Brix \% Bagasse.} \times \text{Tons Bagasse}}{100}$$

$$\text{Tons Fiber in Bagasse} = \frac{\text{Fiber \% Bagasse.} \times \text{Tons Bagasse}}{100}$$

$$\text{Tons Moisture in Bagasse} = \frac{\text{Moisture \% Bagasse} \times \text{Tons Bagasse}}{100}$$

Tons Brix in Cane = Tons Brix in MJ + Tons Brix in Bagasse

Tons Pol in Cane	=	Tons Pol in MJ + Tons Pol in Bagasse
Tons Fiber in Cane	=	Tons Fiber in Bagasse
Pol in Cane	=	Pol in MJ % Cane + Pol in Bagasse % Cane
Fiber % Cane	=	$\dfrac{\text{Tons Fiber in Cane} \times 100}{\text{Tons Cane}}$
Filter Cake % Cane	=	$\dfrac{\text{Tons Filter Cake} \times 100}{\text{Tons Cane}}$
Tons Pol in Filter Cake	=	$\dfrac{\text{Tons Filter Cake} \times \text{Pol \% Filter Cake}}{100}$
Tons Pol in Clear Juice	=	Tons Pol in MJ − Tons Pol in Filter Cake
Tons Brix in Clear Juice	=	$\dfrac{\text{Tons Pol in Clear Juice} \times 100}{\text{Purity of Clear Juice}}$
Tons Non-Sugar in MJ	=	Tons Brix in MJ − Tons Pol in MJ
Mill Extraction	=	$\dfrac{\text{Tons Pol in Mixed Juice} \times 100}{\text{Tons Pol in Cane}}$
or	=	$\dfrac{\text{Pol in Mixed Juice \% Cane} \times 100}{\text{Pol \% Cane}}$
Undiluted Juice Extracted in MJ% Cane	=	$\dfrac{\text{Brix \% MJ} \times \text{Mixed Juice \% Cane}}{\text{Brix \% PJ}}$
Undiluted Juice Lost in Bagasse % Fiber	=	$\dfrac{\text{Brix \% Bagasse} \times 10{,}000}{(\text{Brix \% PJ}) \times (\text{Fiber \% Bagasse})}$
Reduced Mill Extraction	=	$\dfrac{(100 - V)}{7}$

where,

Questionnaire of Sugarcane & Quality Control

$$V = \frac{(100 - \text{Mill Extraction})(100 - \text{Fiber \%Cane})}{\text{Fiber \% Cane}}$$

Java Ratio	=	$\dfrac{\text{Pol \% Cane} \times 100}{\text{Pol \% First Expressed Juice}}$

Pol Balance

Pol in Mixed Juice % Cane	=	$\dfrac{\text{Pol \% Mixed Juice} \times \text{Mixed Juice \% Cane}}{100}$
Pol in Bagasse % Cane	=	$\dfrac{\text{Pol \% Bagasse} \times \text{Bagasse \% Cane}}{100}$
Pol % Cane	=	Pol in MJ % Cane + Pol in Bagasse % Cane
Pol in Filter Cake % Cane	=	$\dfrac{\text{Pol \% Filter Cake} \times \text{Filter Cake \% Cane}}{100}$
Pol in Final Molasses	=	$\dfrac{\text{Pol \% Final Molasses} \times \text{Molasses \% Cane}}{100}$
Sugar in Sugar % Cane	=	$\dfrac{\text{Pol \% Sugar} \times \text{Recover \% Cane}}{100}$
Unknown Loss % Cane	=	$\dfrac{1 \times \text{Sugar in Sugar \% Cane}}{100}$
Pol in Clear Juice % Cane	=	Pol in MJ % Cane - Pol in Fil. Cake % Cane

Brix Balance

Brix in Clear Juice % Cane	=	$\dfrac{\text{Pol in Clear Juice \% Cane} \times 100}{\text{Purity of Clear Juice}}$
Brix in Sugar % Cane	=	$\dfrac{\text{Brix \% Sugar} \times \text{Recovery \% Cane}}{100}$

$$\text{Brix \% Sugar} = 100 - \text{Moist \% Sugar}$$

$$\text{Brix in Final Molasses \% Cane} = \frac{\text{Brix \% Molasses} \times \text{Molasses \% Cane}}{100}$$

Non-Sugar Balance

$$\text{Non-Sugar in Clear Juice \% Cane} = \text{Brix in Clear Juice \% Cane} - \text{Pol in Clear Juice \% Cane}$$

$$\text{Non-Sugar in Sugar \% Cane} = \frac{\text{Recovery \% Cane} \times (\text{Brix \% Sugar} - \text{Pol \% Sugar})}{100}$$

$$\text{Non-Sugar in Final Molasses \% Cane} = \frac{(\text{Brix \% Molasses} - \text{Pol \% Molasses}) \times \text{Molasses \% Cane}}{100}$$

Crystal Balance

$$\text{Crystallizable Sugar in Clear Juice} = \text{Pol in Clear Juice \% Cane} \times \text{Factor (ESG)}$$

$$\text{Crystallizable Sugar in Sugar} = \text{Recovery \% Cane} \times \text{Factor (ESG)}$$

$$\text{Expected Recovery} = \frac{\text{Brix in CJ \% Cane} \times (J - M) \times \text{Unknown Loss}}{(100 - M)}$$

$$\text{Expected Molasses Recovery} = \frac{\text{Brix in CJ \% Cane} \times (100 - J) \times 100}{(100 - M) \times \text{Brix \% Molasses}}$$

Where,
- J = Purity of Clear Juice
- M = Purity of Final Molasses

$$\text{Tons Available Sugar} = \frac{\text{Tons Cane} \times \text{Recovery \% Cane}}{100}$$

$$\text{Tons Available Molasses} = \frac{\text{Tons Cane} \times \text{Molasses \% Cane}}{100}$$

$$\text{ERQV. MJ/PJ} = \frac{(1.4 \times \text{pty of MJ} - 40) \times 100}{(1.4 \times \text{pty of PJ} - 40)}$$

Questionnaire of Sugarcane & Quality Control

$$\text{ERQV. LMJ/PJ} = \frac{(1.4 \times \text{pty of LMJ} - 40) \times 100}{(1.4 \times \text{pty of PJ} - 40)}$$

$$\text{Clarification Efficiency} = \frac{\text{Non-Sugar Removed from MJ \% Cane} \times 100}{\text{Non-Sugar in MJ \% Cane}}$$

$$\text{Non-Sugar removed from MJ \% Cane} = \text{Non-Sugar in MJ \% Cane} - \text{Non-Sugar in CJ \% Cane}$$

$$\text{Clarification Factor} = 100 - \text{Clarification Efficiency}$$

$$\text{Evaporation \% Clear Juice} = \frac{(\text{Brix of Syrup} - \text{Brix of CJ}) \times 100}{\text{Brix of Syrup}}$$

$$\text{Non-Sugar in Molasses} = \text{Tons Brix in Molasses} - \text{Tons Pol in Molasses}$$

$$\text{Theoretical Quantity of Molasses} = \frac{\text{Tons Non-Sugar in Clear Juice} \times \text{Tons Molasses}}{\text{Tons Non-Sugar in Molasses}}$$

$$\text{Molasses Actually Produced \% Theoretically} = \frac{\text{Tons Molasses Produced} \times 100}{\text{Tons Theoretical Molasses}}$$

$$\text{Available Sugar in Material} = (\text{Tons Sugar in Material}) - (M/(100-M)) \times \text{Tons Non-Sugar in Material}$$

$$\text{Molasses in Process} = \frac{(\text{Total Brix in Process} - \text{Total Actual Available Sugar}) \times 100}{\text{Average Brix \% Molasses}}$$

$$\text{Brix-Free Cane Water \% Fiber} = \text{Weight of Cane \% Fiber} - \text{Dirt Correction MJ \% Fiber} - \text{Juice in Cane \% Fiber} - 100$$

$$\text{Available Sugar Per Unit Sugar in MJ} = \frac{S(J-M)}{J(S-M)}$$

Where,
S = Purity of Sugar
J = Purity of Mixed Juice
M = Purity of Final Molasses

ESG Factor	=	$\dfrac{100(J - 28.57)}{J(100 - 28.57)}$

Sugar (ESG)	=	$\dfrac{\text{Sugar in Sugar Produce} \times 100(S - M)}{S(100 - M)}$

S = Pty of Sugar Produced

$\quad = \dfrac{\text{Sugar \% Sugar Produced} \times 100}{(100 - \text{Moisture \% Sugar Produced})}$

Boiling House Recovery	=	$\dfrac{\text{Pol in Commercial Sugar \% Cane} \times 100}{\text{Pol in MJ \% Cane}}$

Basic Boiling House Recovery	=	$\dfrac{100(J - 28.57) \times 100}{J(100 - 28.57)}$

Boiling House Recovery (ESG)	=	BHR × ESG Factor

Boiling House Efficiency (Performance)	=	$\dfrac{\text{Actual Boiling House Recovery} \times 100}{\text{Basic Boiling House Recovery}}$

Overall Extraction or Recovery	=	$\dfrac{\text{Sugar in Sugar Produced} \times 100}{\text{Sucrose in Cane}}$

Virtual Purity of Molasses (Mv)	=	$\dfrac{J(1 - r)}{(1 - rj)}$

Q. What is milling control?

ANS. It is control of all operations starting from cane received to juice extraction. It deals with total sugar input (pol in cane) and output (pol in mixed juice and pol in bagasse) of the mill house.

1. It also helps in mill settings by drawing brix curves.
2. It provides mill sanitation data by ERQV methods.
3. It controls various losses during extraction of juice.

4. Milling control provides data related to mill stoppage, crushing rates, etc.
5. It also helps in providing extraction efficiency values on primary mill extraction and total milling tandem.
6. It also helps in assessing efficiency of preparatory devices.

Q. How many types of data or information are used in control?

ANS. Three types of data are used:

1. Measured data (cane crushed, juice flow, water flow, steam consumption, etc.)
2. Analytical data (brix, pol, moisture, temperature, etc.)
3. Calculated data (pol in cane, pol loss, mill extraction, RME, fiber % cane, java ratio, imbibition % fiber, etc.)

Q. What is imbibition and how many types does it occur in?

ANS. It is the process by which water juice is added on bagasse to mix and dilute the juice for better extraction of pol. It is of two types: simple and compound.

Q. What is simple imbibition?

ANS. If only hot or cold water is added before the last mill, it is known as simple imbibition.

Q. What is compound imbibition?

ANS. Hot water (80 °C) is added before the last mill and the juices are added after the first mill; second mill of four milling tandem by sprinkling on bagasse.

Q. What are some standard values of milling control?

ANS.

Table 18: Some standard values of milling control.

SN	Particular	Unit of Measurement	Standard Value
1.	Pol in Cane	%	+ 12.0
2.	Fiber % Cane	%	13.00 to 15.00
3.	Preparatory Index	%	+ 85.0
4.	ERQV – MJ/PJ	%	98
5.	ERQV – LMJ/PJ	%	85
6.	Imbibition % Fiber	%	250 -300
7.	Primary Extraction	%	70
8.	Mill Extraction	%	94 to 95
9	Pol % Bagasse	%	1.4 to 1.85
10.	Bagasse Moisture	%	< 50.00

Q. What is the effect of extraneous matter on cane processing?

ANS. Extraneous matter such as trash-binding material creates load on clarification and causes color problem.

Q. Why is temperature correction applied in the brix measurement?

ANS. The Brix Spindle is calibrated at 27.5 °C. Hence, if the temperature rises, you have to add and if temperature falls, you have to subtract.

Q. Why is bagasse pol observed in a 400 mm pol tube?

ANS. Bagasse solution is a semi-normal solution. Hence, pol reading is observed in a 400 mm pol tube.

Normal solution = 26 gm of sugar dissolved in a 100 ml DW.

Semi-normal solution = 13 gm sugar dissolved in a 100 ml DW.

Q. Why is moisture factor applied for bagasse pol percentage analysis?

ANS. Initially, bagasse contains some moisture so the moisture factor is applied.

Q. Why is imbibition needed?

ANS. In the milling tandem PI of 85%, 65–70% of juice extraction is done by first mill. After extracting juice from first mill, the remaining material (fiber) gives more resistance for extraction by simple compressing of rollers; hence, imbibition is a necessary part of juice extraction.

Q. Why is cold water not to be used for imbibition?

ANS. Juice extraction is not proper if the temperature of water used does not soften the hard rind of the cane. The material of the cell walls, which is permeable only by osmosis would soften, and water would thus have direct access to the juice contained in the cell.

Q. What should the range of imbibition be?

ANS. The range should be 30–40% on cane and added water percentage fiber should be 250–275% on fiber.

Q. What is the difference between ME and RME?

ANS. The main difference between ME and RME is that RME is based on 12.5% fiber whereas ME is based on juice extraction.

Q. What is the difference between RME_{Deer} and RME_{Mittal}?

ANS. The mill extraction reduced on common 12.5% fiber basis is called RME_{Mittal}. This is on pol basis whereas mill extraction reduced on varying fiber percentage cane is called RME_{Deer}; this is on juice basis.

Q. What is inversion and how can it be controlled?

ANS. Conversion of sucrose into glucose and fructose is known as inversion.

Inversion can be controlled by:

1. Uniform crushing rate.
2. Appropriate retention time.
3. Maintaining the temperature of juice in I vessel.

Q. Why is mill sanitation used?

ANS. To avoid the following:

1. Acid inversion: 22%
2. Enzyme inversion: 16%
3. Microbial inversion: 7–8%

Q. How can pol percentage bagasse and moisture percentage bagasse be measured?

ANS. For pol percentage bagasse, take 250 gm bagasse in Rapi pol extractor and add 2,000 ml of water. Rotate for about fifteen minutes. After fifteen minutes, remove the liquid from the Rapi pol extractor and add lead sub acetate and filter by the usual method. See the pol reading in the 400 mm pol tube. The reading gives the direct pol percentage.

Pol percentage bagasse (400 mm) = pol reading × moisture factor.

For moisture percentage bagasse, take 100 gm of the bagasse in a weighted tray perforated on all sides. Dry in an oven at 110 °C. Carry out the drying to constant weight. The loss in weight gives the moisture percentage bagasse.

Q. How can Preparatory Index (PI) be determined?

ANS. Take 4–5 kg of prepared cane after it has been through the fibrizer.

Take 500 gm prepared cane and add 5 liters of water in a ball mill. Rotate for about forty minutes and take pol reading, which is P1.

Take another 1,000 gm of prepared cane and add 3 liters of water in a Rapi pol extractor and rotate for about half an hour till the brix of the solution is constant. Take pol reading, which is P2.

Now calculate using the given formula:

$$r = P1/P2$$

$PI = 1{,}000 \times r / (3.888 - 0.838\, r)$

Q. How can brix measurement of juices be taken?

ANS. Fill the juice to overflow in a cylinder that should be vertical. Allow the air to escape by letting it stand for twenty minutes. Gradually lower the standardized Brix Spindle of the approximate range into position. When the spindle becomes steady take the reading, keeping your eyes in line with the plane surface of the liquid. Note the temperature of the juice from the thermometer. The brix of the juice is given by the reading of hydrometer spindle + hydrometer correction. If any temperature correction as read from the tables IV-A and IV-B corresponding to the recorded temperature.

Q. How can pol measurement of juices be taken?

ANS. Transfer about 200 ml juice into a 300 ml flask, which should have been rinsed with the juice before transferring. Add 2–3 gm lead sub acetate. Shake the flask well and filter. Use dry filter paper and dry funnel. Cover the funnel with a watch glass during filtration. Collect the filtrate in a clean dry beaker. Reject the first few ml of the filtrate. Rinse a 200 ml pol tube three times with a portion of the filtrate. Fill the tube, ensuring that there are no air bubbles in it. Do not tighten the end caps of the pol tubes. Then take the pol reading in a saccharimeter. Take the average of five readings as the pol reading. Then find the pol percentage juice by referring to Schmitz's Table (V-A and V-B).

Q. What is dextran and how can it be determined in cane juice?

ANS. Dextran is defined as a high molecular weight and predominantly straight-chained glucose polymer with a majority of α −(1–6) glucosidic linkage, formed by the action of certain species of bacteria, especially *leuconostoc mesenteroides*, on sucrose.

Determination of Dextran in Cane Juice: Take 25 ml of cane juice. Add 25 ml of 10% TCA. Now add 0.5 gm supercel. Filter the solution through filter paper. Take 5 ml of filtrate and add 5 ml absolute alcohol. Wait for twenty minutes. Measure the absorbance at 720 nm on sucroscan.

For blank sample: Take 5 ml filtrate and 5 ml of distilled water and compare with the standard graph, which has to be prepared by the known concentration using sucroscan.

Q. Which ICUMSA method is used for dextran measurement?
ANS. GS 1/2/9–15 (2007).

Q. What is ICUMSA?
ANS. ICUMSA brings together the activities of the National Committees for sugar analysis. The SOP for analyzing plantation white sugar, raw sugar, refined sugar and other sugar house products have been recommended by ICUMSA.

Q. What does GS mean and what is its indication for different products?
ANS. GS means General Specified. For different sugar products, the following are used:

1. Raw Sugar - GS-1
2. White Sugar - GS-2
3. Specialty Sugars and Refined Liquid Products - GS-3

4. Molasses - GS-4
5. Sugar Cane - GS-5
6. Cane Sugar Processing - GS-7

Q. What is color?

ANS. Color does not mean the visual appearance of sugar but the amount and intensity of coloring matter or colorants present per unit amount of sugar. A pure sugar crystal is colorless.

The color of sugar is measured in solution.

Color is specified as ICUMSA Units (IU) which is the absorbency index of a filtered sugar solution at 420 nm, multiplied by 1,000.

Q. Why is commercial sugar colored?

ANS. Commercial white sugar crystals are colored due to the presence of impurities as inclusions within the sugar crystals or as a thin film of molasses on the surface of the crystal.

Q. What is the method used for different sugars and sugar products?

ANS. The method for a particular sugar or intermediate product is selected according to the scope and application range of the method.

Table 19: Sugar methods and scope.

Method No. and Year	Status	Scope	Color Range
GS 2/3–10 (2002)	Official	White sugar	Below 50 IU
GS 2/3–9 (2002)	Accepted	All crystalline sugars, powdered white sugar, very pure syrups, white sugars, plantation white sugar	Up to 600 IU with no lower limit
GS 1/3–7 (2002)	Official	Raw sugar, partly refined sugar, brown sugar, juices, syrups	From 250 IU with no upper limit
GS 9/1/2/3–8 (2004)	Official	White sugar, plantation white sugar, raw sugar	Up to 16,000 IU with no lower limit

For all methods, the basic principles of color measurement are the same.

Q. What is the cell length for sugar color measurement?

ANS. Cell length for sugar color measurement should be selected depending upon the color range of the product.

For low-colored sugars, higher cell length is used whereas higher color products are analyzed with lower cell length.

There are two types of cell length. One is 1 cm and another is 5 cm and 10 cm.

1. 1 cm cell length — for high-colored products (raw sugar, syrups, juices)
2. 5 cm cell length — for plantation white sugar and refined sugars
3. 10 cm cell length — preferred for refined sugar

Q. What is the formula for sugar color measurement?

ANS. The formula for sugar color measurement is

$$\text{ICUMS Color} = A_s \times 1{,}000/bc$$

Where,

As	=	Absorbance at 420 nm
b	=	Cell length in cm
c	=	Concentration in gm/ml

Q. What is the dilution for different process intermediate products?

ANS.

Table 20: Dilution for different process intermediate products.

Intermediate products	Dilution
Primary Juice, Mixed Juice, B Massecuite, C Massecuite, A-h = Heavy, B Heavy, C Light, Final Molasses, Filtrate Juice	Brix 2.5 +/- 0.5
Sulphited Juice, Clear Juice, Syrup, Sulphited Syrup, A-Light, A Massecuite, Melt	Brix 5.0 +/- 1.0
B Sugar, CFW Sugar, CAW Sugar, Dry Seed	Brix 20.0 +/- 5.0

Q. How can dextran formation in milling be controlled?

ANS.

A. It can be controlled physically by:
 1. Regular washing of mills with hot water/steam.
 2. Removal of biofilm.
 3. About 60% of the problem is solved by physical cleaning.

B. It can be controlled chemically by:
 1. Use of biocides/mill sanitation chemicals.
 2. Use of dithiocarbamate compounds which act on enzyme systems.
 3. Use of quaternary ammonium compounds which act on the cell wall.

Q. What are favorable conditions for dextran formation at mills?

ANS. The favorable conditions for dextran formation at mills are:
 1. Stagnation of bagasse.
 2. Formation of biofilm.
 3. Inaccessible cleaning areas.
 4. Pockets of slimy growth.
 5. Use of gunny bags to avoid juice splashing.

Q. What should be the clear juice pH? Why must this be maintained?

ANS. The pH of the clear juice should be maintained in the range of 6.8–7.1. Above this range, the destruction of reducing sugar takes place producing a dark coloration to the juice. At the same time, if lime is added to increase pH, lime remains in the solution form creating turbidity in the juice as it remains in colloidal form. If pH goes below 6.8, inversion of sucrose takes place.

$$C_{12}H_{22}O_{11} + H_2O \longrightarrow C_6H_{12}O_6 + C_6H_{12}O_6$$
$$\text{Sucrose} \quad \text{water} \qquad\qquad \text{glucose} \quad \text{fructose}$$

Q. What is clarification factor and clarification efficiency?

ANS. Clarification efficiency is the ratio of non-sugar removed percentage cane to the non-sugar in mixed juice percentage cane.

$$\text{Clarification efficiency} = \frac{\text{Non-Sugar Removed \% Cane} \times 100}{\text{Non-Sugar in Mixed Juice \% Cane}}$$

Non-Sugar Removed = Non-Sugar in MJ % Cane - Non-Sugar in C.J % Cane

Clarification efficiency for carbonation process is in the range of 20–25% whereas for sulphitation process it is in the range of 8–15% that indicates good working efficiency of clarifier.

Clarification factor = 100 - clarification efficiency.

Q. What is the difference in conductivity ash and sulphited ash percentage?

ANS. During the determination of conductivity ash there is no use of any chemical whereas in the determination of sulfated ash concentrated sulphuric acid is used.

Q. What is the effect of lime quality on juice clarification?

ANS. Lime is the important basic chemical used in clarification. If the quality of the lime is poor, it affects the quality of clear juice. If lime quality is poor, heavy scaling is seen in evaporator bodies and consumption of lime increases.

Normal range: 0.16–0.18% cane is the consumption of lime.

Q. How many ICUMSA methods of sugar analysis are used in India?

ANS.

1. Method GS-2/3–9 (1994) color of sugar (TEA/HCL).
2. Method GS-2/3–7.
3. Method GS-2/3–8.
4. Method GS-2/3–10 (color of sugar, distilled water method).

Q. What is the grading of molasses?

ANS.

TRS is above 50% A Grade.

TRS is above 45–50% B Grade.

TRS is above 40–45% C Grade.

Q. How can safety factor be calculated?

ANS. Safety factor = $\dfrac{\text{Moisture \%}}{100 - \text{Pol \%}} = \dfrac{\text{Moisture \%}}{\text{Non-Sugars}}$

Q. What are unknown losses?

ANS. Losses due to spillage, leakage, entrainment and inversion are unknown losses.

Q. What is meant by MA and CV?

ANS. MA = Mean Apparatus.

CV = Coefficient of Variation [35 in maximum 30 means good crystal variation].

Q. What is the percentage of CaO in different limes?

ANS. 1. Limestone: 55–56%

2. Quick lime: 85–92%

3. Hydrated lime: 73–75%

4. Burnt lime: 70–75%

Q. What are the objectives of clarification?

ANS. The objectives of the clarification of mixed juice are:

1. Removal of maximum non-sugars.
2. Elimination of suspended and colloidal impurities.
3. Removal of color-forming compounds.
4. Obtaining brilliant, light colored, transparent clear juice free from any suspended impurities.

Q. Which clarification processes are followed for production of different sugars?

ANS.

1. Defecation process	— Raw sugar production.
2. Carbonation process	— Refined sugar.
3. Sulphitation process	— Plantation white sugar.
4. Defco-Melt Phosphotation process	— Refined sugar.

Q. What is the pol balance in a sugar factory?

ANS. Pol in cane - Recovery % cane = total losses

Total losses = Pol in bagasse + Pol in final molasses % cane + Pol in filter cake % cane + Unknown losses.

Q. What are the different types of reports prepared by the lab in-charge?

ANS. The different types of reports prepared by the lab in-charge are:
1. Daily manufacturing report.
2. Weekly report.
3. Fortnightly report.
4. Monthly report–RT 7C.
5. Final Report–RT 8C.

Q. What is the difference between plantation white sugar and raw sugar manufacturing?

ANS. Sulphitation is not done in raw sugar manufacture whereas in plantation white sugar sulphitation is the most important step. In centrifugal a massecuite, three washes are undertaken in plantation white sugar whereas a single wash is done for raw sugar.

QUALITY MANAGEMENT SYSTEM

In this chapter, general idea about the Quality and Quality management system and how can implement, internal audit, new standard clauses etc. for all professionals

Q. What is Quality Policy?

ANS. Quality policy is a policy related to quality.

Q. What is a policy?

ANS. A policy are intentions and directions of an organization as formally expressed by top management.

Q. What are the requirements for a quality policy?

ANS. The quality policy is compatible with the strategic direction and the context of organization. The quality policy is communicated, understood and applied within organization.

Q. What is the relationship between quality policy and top management?

ANS. The quality policy is established, reviewed and maintained by Top Management.

Q. What about the completeness of the quality policy?

ANS. The quality policy includes a commitment to satisfy applicable requirement for products & services.

Q. What is the relationship between quality policy and quality objectives?

ANS. The quality policy provides a framework for setting and reviewing quality objectives. Quality objectives are consistent with the quality policy.

Q. What about the appropriateness of the quality policy?

ANS. The quality policy is appropriate to the purpose and context of the organization.

Q. What about the availability of the Quality policy?

ANS. The quality policy is available as documented information and available to relevant interested parties.

Q. What is the relationship between quality policy and persons?

ANS. Persons doing work under the organization control are aware of the quality policy.

Q. What is quality assurance?

ANS. The processes that ensure production quality meets the requirements of customers. Quality assurance is about putting procedures and processes in place which aim to "assure" quality is achieved rather than relying on checking that it has.

Q. What is quality?

ANS. Quality is about meeting the needs and expectations of customers. If a product or service meets all those needs and expectations, then it passes the quality test.

ISO 9000 defines quality simply as "the degree to which a set of inherent characteristics fulfils requirements

Q. What is quality control?

ANS. The process of inspecting or checking products to ensure that they meet the required quality standards, which can be performed during the production process or by checking the quality of finished goods & services.

Q. What is the impact of poor quality?

ANS. It has the following impact:
1. Loss of customers (expensive to replace and they may tell others about their bad experience),
2. Cost of reworking or remaking product,
3. Cost of replacements or refunds,
4. Waste of materials, and
5. Damage to reputation in the marketplace.

Q. What are the factors used to assess quality?

ANS. Performance (fit for purpose), appearance, availability and delivery, reliability/durability, price/value for money, functions, after-sales service, image and reputation.

Q. How can one judge if the product has failed?

ANS. The product is considered to have failed if:
1. There is a breakdown or unexpected wear and tear.
2. The product does not perform as promised.
3. The product is delivered late.
4. Instructions/directions for use are poor.
5. Customer service is unresponsive.
6. The product is poorly packaged or appears damaged.
7. The product develops a bad reputation.

Q. What is QC/QC activity?

ANS. QC stands for quality control. It is an activity to improve the quality of products undertaken by employees in the field by considering management and improvement of manufacturing processes in a manufacturing industry.

Q. What are seven QC tools?

ANS. The seven QC tools are as under:
1. Pareto chart,
2. Cause and effect diagram,
3. Check sheet,
4. Graph,
5. Histogram,
6. Scatter diagram, and
7. Control chart.

Q. What is TQM?

ANS. Total Quality Management (TQM) is a management philosophy committed to a focus on continuous improvements in products and services with the involvement of the entire workforce. The whole business understands the need for quality and seeks to achieve it. Everyone in the workforce is concerned with quality at every stage of the production process. Quality is ensured by workers and not inspectors.

Q. What is QI?

ANS. QI is Quality Improvement. The many benefits of a functional QI are:

1. It helps to overcome challenges from free trade and globalization.
2. It enables access to international markets and preserves domestic markets.
3. It promotes innovation and competitiveness.
4. It protects consumers.
5. It assists regulators and service providers.
6. It advances economic development.
7. It encourages regional integration.

Q. What is quality management?

ANS. The term 'quality management' (QM) is defined in ISO 9000 as "coordinated activities to direct and control an organization with regard to quality." To direct and control an organization, its management should first set out its quality policy and related quality objectives and then specify activities related to quality planning, quality control, quality assurance and quality improvement. The objective of QM is to ensure that all company-wide activities necessary for enhancing the satisfaction of customers and other stakeholders are carried out effectively and efficiently. QM focuses not only on product/service quality but also on the means for achieving it.

Q. What are the components of quality management?

ANS.

Fig.24: The four components of quality management.

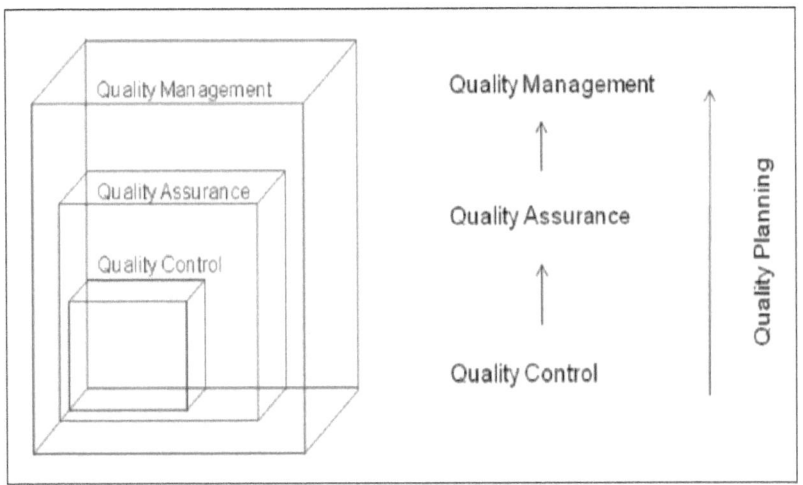

- Quality Planning (QP): Can we make it OK?
- Quality Control (QC): Are we making it OK?
- Quality Assurance (QA): Will we continue making it OK?
- Quality Improvement (QI): Could we make it better?

Q. What is Quality Planning (QP)?

ANS. It is "a part of quality management focused on setting quality objectives and specifying necessary operational processes and related resources to fulfil quality objectives." (ISO 9000:2005 3.2.9) QP is a systematic process that translates quality policy into measurable objectives and requirements and lays down a sequence of steps for realizing them within a specified timeframe. The results of QP are presented, for use by all concerned, in the form of a quality plan, a document specifying which procedures and associated resources will be applied by whom and when. Such quality plans are prepared separately for specific processes, products or contracts.

Q. What is Quality Control (QC)?

ANS. Quality Control is "a part of quality management focused on fulfilling quality requirements." (ISO 9000:2005 3.2.10) QC helps in evaluating the actual operating performance of the process and product and, after comparing actual performance with planned targets, it prompts action on deviations found, if any. QC is a shop-floor and online activity that requires adequate resources, including skilled people, firstly to control the processes and then to carry out timely corrections when process and/or product parameters go beyond prescribed limits.

Q. What is Quality Assurance (QA)?

ANS. Quality Assurance is "a part of quality management focused on providing confidence that quality requirements will be fulfilled." (ISO 9000:2005 3.2.11) Both customers and management need an assurance of quality because they are not in a position to oversee operations themselves. QA activities establish the extent to which quality will be, is being or has been fulfilled. The means to provide the assurance needs to be built into the process, such as documenting control plans and specifications, defining responsibilities, providing resources, performing quality audits, maintaining records and reporting reviews. QA is more comprehensive than QC, which is part of it.

Q. What is Quality Improvement (QI)?

ANS. It is "a part of quality management focused on increasing the ability to fulfil quality requirements." (ISO 9000:2005 3.2.12) To maintain performance and position in the market, QI activities will have to be carried out on a continual basis. Such improvement activities include refining existing methods and modifying processes to reduce variations and to increase yield while consuming fewer resources.

Q. What does "ISO" stand for?

ANS. ISO is not an acronym; it is a word chosen by the International Organization for Standardization. "ISO" is taken from the Greek word "isos," meaning equal. The three official languages of ISO are English, French and Russian; thus, the organization's name would have different acronyms in different languages. For this reason, it adopted the short name ISO (a registered trademark of the organization) which is the same in every country.

Q. What is the International Organization for Standardization?

ANS. The organization today known as ISO began in 1926 as the International Federation of the National Standardizing Associations and became known as the International Organization for Standardization in 1947. It is a worldwide federation of national standards bodies from more than 160 countries, one from each country. The national standards bodies make up the ISO membership and they represent ISO within their country. The organization's mission is to promote the development of standardization to facilitate the international exchange of goods and services, and to develop cooperation in the spheres of intellectual, scientific, technological and economic activity. Its work results in international agreements, which are published as international standards.

Q. What are the member categories of ISO?

ANS. There are three membership categories for national standards bodies.

1. Full members or member bodies influence the development and strategy of ISO standards by participating and voting in ISO technical and policy meetings. Full members sell and adopt ISO International Standards nationally.
2. Correspondent members observe the development of ISO standards and strategy by attending ISO technical and policy meetings as observers. Correspondent members can sell and adopt ISO International Standards nationally.
3. Subscriber members keep up to date on ISO's work but cannot participate in it. They do not sell or adopt ISO International Standards nationally.

Q. What is ISO 9000?

ANS. ISO 9000 family of standards represents an international consensus on good management practices with the aim of ensuring that the organization can repeatedly deliver products or services that meet the client's quality requirements. These good practices have been distilled into a set of standardized requirements for a quality management system, regardless of what the organization does, its size, or whether it is in the private or public sector. The family of ISO 9000 standards have been developed by ISO and is made up of four core standards:

a) ISO 9000:2005—Fundamentals and Vocabulary
b) ISO 9001:2008—Quality Management Systems–Requirements

c) ISO 9004:2009—Quality Management Systems Guidelines for Performance Improvements
   d) ISO 19011:2011—Guidelines for Quality and/or Environmental Management Systems Auditing.

Q. What is the role of BIS in ISO 9000?

ANS. BIS is the National Standards Body of India and is a founder member of ISO. BIS represents India in ISO. The ISO Technical Committee (TC) number 176 (ISO/TC 176) and its sub-committees are responsible for the development of ISO 9000 standards. Quality and industry experts from India including BIS officers nominated by BIS participate in the meetings of the ISO/TC 176 and its sub-committees.

Q. What is the difference between ISO 9000 standards and IS/ISO 9000 standards?

ANS. There is no difference; they are exactly the same. BIS has adopted the above mentioned ISO 9000 standards and these are numbered as IS/ISO 9000:2005; IS/ISO 9001:2008; IS/ISO 9004:2009 and IS/ISO 19011:2011. These standards published by the BIS are an exact replica of ISO 9000 standards. BIS also provides certification against IS/ISO 9001:2008 under its Management Systems Certification activity.

Q. What is IS/ISO 9004:2009?

ANS. The requirements of IS/ISO 9004:2009 should be implemented by organizations who intend to further improve beyond the requirements of IS/ISO 9001:2008. With an important element of "self-evaluation," IS/ISO 9004:2009 is not amenable to certification.

Q. What are the main benefits of implementing IS/ISO 9001:2008 and Quality Management System Requirements?

ANS.

1. It provides an opportunity to increase value to the activities of the organization.
2. It improves the performance of processes/activities continually.
3. It will lead to satisfaction among customers.
4. It helps attend to resource management.

5. It involves implementation of statutory and regulatory requirements related to products/services.
6. It results in better management control.

Q. What is the difference between a certification body and a registration body?

ANS. The term "certification body" is used in some countries, like India, because BIS as a certification body issues certificates (licenses). Elsewhere, registration body is preferred since it "registers" organizations complying with ISO 9000.

Q. What is accreditation?

ANS. In simple terms, accreditation is certification of the certification body. "Accreditation" should not be used interchangeably with certification or registration.

Q. What is a "standard"?

ANS. A standard is a document that provides requirements, specifications, guidelines or characteristics that can be used consistently to ensure that materials, products, processes and services are fit for their purpose. ISO international standards ensure that products and services are safe, reliable and of good quality. They are strategic tools that reduce costs by minimizing waste and errors and increasing productivity. They help companies access new markets and facilitate free and fair global trade. Government and industries around the world have been using international standards for more than half a century to facilitate trade, establish a technical base for regulation and safeguard consumers.

Q. What is a quality management system?

ANS. A Quality Management System (QMS) consists of policies, procedures, SOPs and records that provide proof of goals, assign responsibility, describe how those responsibilities are to be performed and provide evidence of past accounts or occurrences of compliance. QMS captures the requirements of an organization and structurally provides a roadmap that explains who, what, when, where and how sustainable and repeatable outcomes will be achieved. QMS is a set of interrelated activities or processes carried out within an organization to provide a product or service that satisfies customer requirements and expectations.

Q. What is ISO 9001?

ANS. ISO 9001 is an international standard that provides the framework for an effective QMS. It is one of the most widely recognized and commonly used standards throughout the world, and applies to all types of businesses regardless of their size, industry, product or service offering. An ISO 9001 QMS is intended to provide customers assurance in the quality of products or services.

Q. What is ISO 9001 certification?

ANS. ISO 9001 certification is a seal of quality that indicates an organization's QMS meets the ISO 9001 requirements. It is based on an independent evaluation or audit performed by a registrar who provides written assurance that the ISO QMS standard has been met. ISO certificates are issued for three years with the requirement that surveillance audits be performed on an annual basis. A company that is ISO 9001 conforming meets ISO's QMS requirements but has not been formally certified by an independent registrar.

Q. Why must ISO 9001 be implemented?

ANS. ISO 9001 should be implemented to:

1. Improve financial performance;
2. Reduce operational costs through improved process effectiveness and efficiency;
3. Improve customer satisfaction;
4. Improve product and service quality, including on-time delivery;
5. Reduce defects and product recalls;
6. Improve employee satisfaction;
7. Motivate employees by clarifying their roles, responsibilities and contributions to the organization's objectives.

Q. How does an organization become ISO 9001 certified?

ANS. The following steps are required for ISO 9001 certification:

1. Develop a thorough understanding of the ISO 9001 standard.
2. Perform a gap assessment to identify where current systems meet the standards and where they do not.
3. Develop an action plan and timeline for achieving certification.
4. Select a registrar.
5. Develop a QMS that defines existing processes and procedures for doing business and conforms to the standard.
6. Define the organization's purpose.
7. Define the quality policy and objectives of the organization.
8. Identify and document processes that are needed to produce the product/service.
9. Develop a Quality Manual.
10. Develop documented procedures as required by the standard.
11. Implement the QMS.
12. Communicate and provide training.
13. Perform internal audit.
14. Conduct management review.
15. Have certification audit performed by the registrar. The quality system should be in place two-three months prior to the audit.

Q. What is meant by quality systems?

ANS. These are standard procedures for carrying out quality control and cover the whole life of a product, that is, design, procurement, manufacture and operation. Quality systems also cover the documentation needed to ensure compliance with set quality standards and permissible deviation criteria.

Q. What are new quality management principles as per ISO 9001:2015?

ANS. The seven new quality management principles are as under:

1. Customer focus,
2. Leadership,
3. Engagement of people,
4. Process approach,
5. Improvement,

6. Evidence-based decision making, and
7. Relationship management.

Q. How many clauses does the ISO 9001:2015 have?

ANS. ISO 9001:2015 have 10 main clauses as under (sub-clauses not included):

1. Scope.
2. Normative Reference.
3. Terms and Definition.
4. Context of the Organization (basic understanding about the organization, expectations of interested parties, scope of QMS, processes of QMS).
5. Leadership (commitment by the top management for QMS implementation, policy, customer focus, roles, responsibilities and authorities).
6. Planning for QMS (risks and opportunities, objectives, changes to QMS).
7. Support (people, infrastructure, process environment, monitoring and measuring resources, knowledge, competence, awareness, communication, documented information—creation, updates and control).
8. Operation (planning and control, determining the requirements for products and services, design and development, control of externally provided products and services, production and service provision, release of products and services, control of non-conforming products and services).
9. Performance Evaluation (monitoring, measuring, analyzing and evaluating, internal audit, management review).
10. Improvement (non-conformity and corrective action, continual improvement).

Q. Why has ISO 9001 been revised?

ANS. ISO standards are reviewed every five years to ensure that they are still relevant in the marketplace. The last major revision of ISO 9001 took place in 2000 and the world has altered significantly since then. To remain a tool that consistently helps organizations maintain high standards, ISO 9001 must fully consider modern industry needs. The new revised standard is intended to reflect this.

Q. Who was involved in the revision?

ANS. Experts from around 50 countries were actively involved in drafting ISO 9001:2015. They include representatives from a broad range of stakeholders, from small businesses to multi-nationals, government departments to industry and trade associations. This is to ensure that the standard reflects the broadest stakeholder representation possible.

Q. What are the requirements of ISO?

ANS. ISO requirements initially involve utilizing the standard to write a quality policy, quality manual and quality objectives, and then using the process approach to address the other requirements of the standard. The conventional ISO saying is:

1. Document what you do,
2. Establish a process for the service,
3. Perform to your documentation,
4. Provide the service based on the process,
5. Record the results of your work,
6. Appropriately maintain all recorded information,
7. Audit the documentation for effectiveness ("audit effectiveness"), and
8. Audit using the process approach.

Q. What is the minimum requirement for a Quality Control System?
ANS.

1. Provide adequate infrastructure.
2. Establish a proper work environment and hygiene.
3. Make workers aware of quality practices.
4. Make available easy-to-understand raw material and product specifications.
5. Make available easy-to-understand instructions for performing work.

6. Purchase raw materials of acceptable quality from suppliers.
7. Store raw materials and other supplies properly to prevent mix up and spoilage.
8. Check or inspect raw materials and other supplies before use.
9. Maintain machines, building and production facilities regularly.
10. Maintain measuring instruments and check their accuracy (calibration).
11. Follow proper production process steps.
12. Control the process to achieve product specification.
13. Prevent the manufacture of defective products.
14. Make use of statistical techniques, such as sampling plans and control charts for process control and other QC tools.
15. Conduct stage inspection during manufacturing.
16. Conduct final inspection of the finished product and packaging against set specifications.
17. Handle the product with care both during internal processing and during delivery to the customer.
18. Obtain feedback on the findings of the final inspection and take appropriate action.
19. Analyze customer complaints or feedback and take action to remove the causes of complaints.
20. Take corrective action on deviations found, if any.

Q. How can an ISO 9001 quality management system be set up?

ANS. The following steps may be used to set up an ISO 9001 quality management system:

Step 1: Team nomination.

Step 2: Gap analysis.

Step 3: Documentation.

Step 4: Training and implementation.

Step 5: Internal audit and improvement.

Step 6: Management review.

Step 7: Certification.

Q. What are the objectives of quality?

ANS. Quality objectives should be SMART.

Specific: relevant to the process or task to which they are being applied.

Measurable: expressed in terms that can be measured using available technology.

Achievable: within the resources that can be made available.

Realistic: in the context of the current and projected work load.

Timely: specific start and completion dates.

Q. What is the history of ISO 9001 QMS? When was it revised?

ANS. Standard ISO 9001 was published in 1987. It underwent several revisions—in 1994, 2000 and 2008. In the beginning it was oriented only for production. When service-providing companies had problems with the application, the standard was unified and universalized. ISO 9001 standard has been revised and published on September 2015.

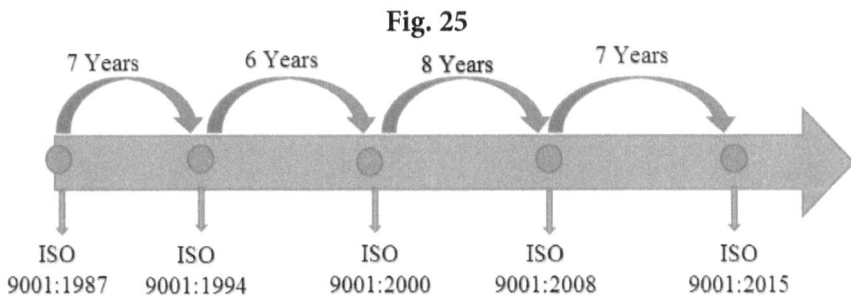

Fig. 25

Q. What is the new model of a process-based QMS?

ANS.

Fig. 26: Model of a process-based QMS.

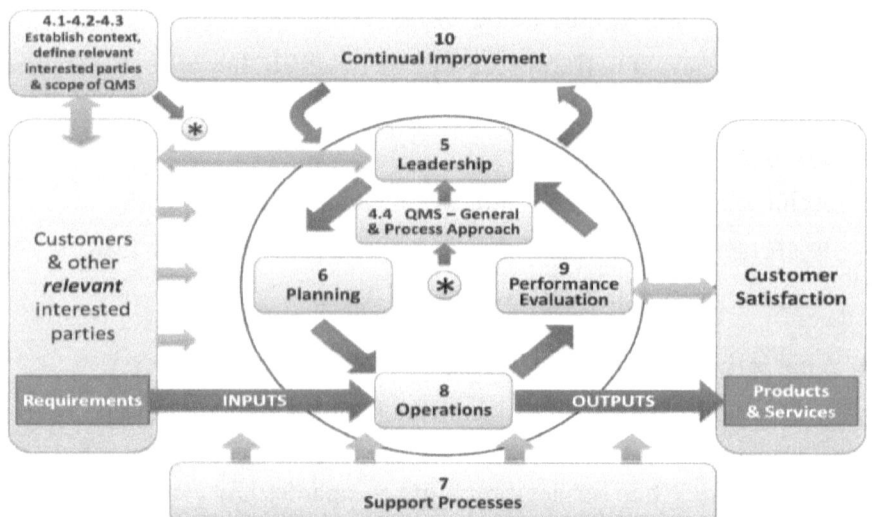

Q. What does a single process within the system entail?

ANS.

Fig. 27: A single process within the system.

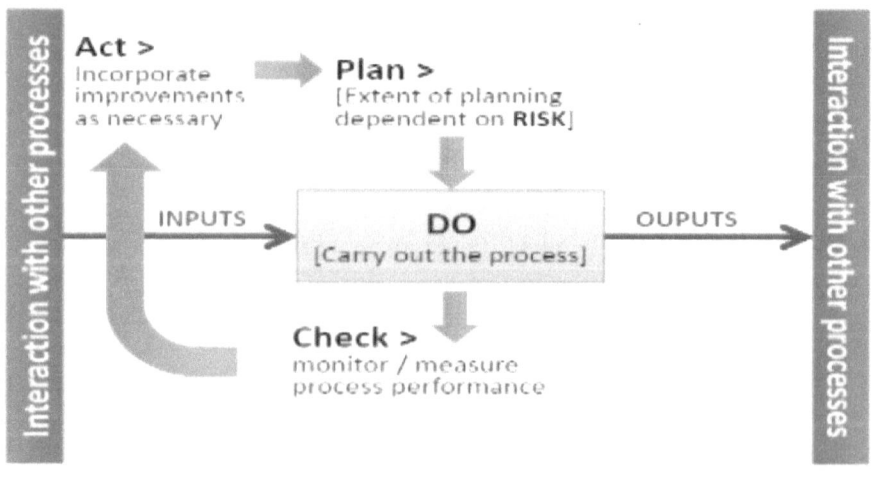

Fig. 28: A group of interrelated activities and related resources that transform inputs into outputs.

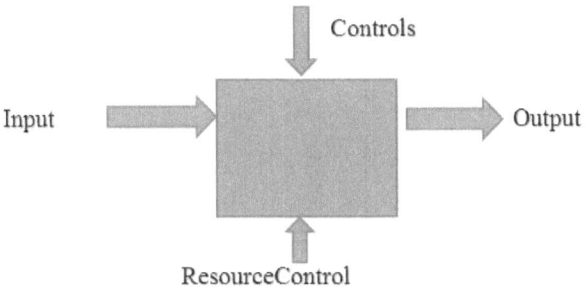

Q. What is an internal quality audit?

ANS. An internal quality audit is one that is performed by or at the direction of members of the organization.

Q. What are the principles of auditing?

ANS. The principles of auditing are:

1. Ethical conduct,
2. Fair presentation,
3. Professional care,
4. Independence,
5. Objectivity,
6. Impartiality,
7. Evaluations based on evidence,
8. Competence,
9. Cooperation, and
10. Trust.

Q. How many types of audit are there?

ANS. There are three types of audits:

1. External independent audits—third party.
2. Customer audits of suppliers—second party.
3. Internal audits—first party.

Q. How can an organization prepare for audit?

ANS. An organization can prepare for an audit by:

1. Defining audit objectives,
2. Defining audit scope,
3. Defining audit resources,
4. Defining audit criteria,
5. Preparing and distributing an audit notification to the auditee,
6. Gathering and understanding relevant documents, and
7. Preparing work plan or audit plan.

Q. What are the criteria for internal and external audit?

ANS. Internal audit criteria are:

1. Standard operating procedures,
2. Quality system procedures,
3. Training procedures,
4. Calibration procedures,
5. Emergency procedures,
6. Records procedures,
7. Customer complaints procedures,
8. Maintenance procedures, and
9. Product specification procedures, etc.

External audit criteria are:

1. ISO standard,
2. Sector-specific documents,
3. Government regulations and industry codes,
4. Corporate policy,
5. Customer requirements, reflected in the contract and purchasing specifications and
6. Market and customer requirements for better products, improved services, or lower prices that have been accepted by senior management as internal goals or requirements.

Q. What is the agenda of an internal audit?

ANS. An internal audit has the following agenda:

1. Plan, establish, implement and maintain audit programs.
2. Determine frequency, methods, responsibilities, planning requirements and reporting.
3. Define the criteria and scope for each audit.
4. Select auditors and conduct audits to ensure objectivity and impartiality of the audit process.
5. Ensure that the results of the audits are reported to relevant management.
6. Take appropriate corrective action without undue delay.
7. Retain documented information as evidence of the implementation of the audit program and the audit results.

Q. What is conformity assessment?

ANS. Conformity assessment is a collective term covering the many elements required to demonstrate that a product or a service complies with stated technical and other requirements. In general, testing, inspection and certification are considered core conformity assessment services and they are used either individually or collectively as circumstances demand.

Q. What is management system certification?

ANS. Management system certification deals with the processes and procedures of the manufacturer, producer, supplier or service provider. The management system can be assessed against the requirements of the relevant standards and, if found to conform, certified by a certification body.

Q. What is the difference between effectiveness and efficacy as per QMS?

ANS. ISO DIS 9001:2015 defines *effectiveness* (3.06) as "extent to which planned activities are realized and planned results are achieved." Efficacy is defined as "the power to produce an effect." They do indeed both refer to the ability of an organization to achieve its objectives.

Q. What does the ISO 9001 certification not mean?

ANS. ISO 9001 defines the requirements for an organization's QMS, not its products. Certification of compliance to ISO 9001 is meant to provide confidence in the organization's ability to consistently provide products that meet customer and applicable statutory and regulatory requirements. It is important to note that it does not necessarily mean that the organization will always achieve 100% product conformity.

Q. What are the changes in QMS clauses?

ANS. The standard is based on Annex SL of the ISO Directives, a high-level structure (HLS) that standardizes sub-clause titles, core text, common terms and core definitions to enhance compatibility and alignment with other ISO management system standards.

The main changes in the new version of ISO 9001:2015 are:

1. The adoption of the HLS as set out in Annex SL of ISO Directives Part One,
2. An explicit requirement for risk-based thinking to support and improve the understanding and application of the process approach,
3. Fewer prescribed requirements,
4. Less emphasis on documents,
5. Improved applicability for services,
6. A requirement to define the boundaries of the QMS,
7. Increased emphasis on organizational context,
8. Increased leadership requirements, and
9. Greater emphasis on achieving desired outcomes to improve customer satisfaction.

Table 21

ISO 9001:2008		ISO 9001:2015	
Clause No.	Particulars	Clause No.	Particulars
1	Scope	1	Scope
2	Normative Reference	2	Normative Reference
3	Terms & Definitions	3	Terms & Definitions
4	Quality Management System	4	Context of the Organization

5	Management Responsibility	5	Leadership
6	Resource Management	6	Planning
7	Product Realization	7	Support
8	Measurement & Improvement	8	Operation
		9	Performance Evaluation
		10	Improvement

Q. What are the reasons for these changes?

ANS. The reasons for these changes are as under:

1. In the past twenty-five years, many other management system standards have come into use worldwide.
2. Organizations that use multiple management system standards are increasingly demanding a common format and language that is aligned across those standards.
3. There is a need to decrease the emphasis on documentation.
4. There has been an increased emphasis on achieving value for the organization and its customers.
5. The emphasis on risk management to achieve objectives has increased.

Q. Which standard is easier to implement, the 2008 version or the 2015 version?

ANS. The main advantage of the 2015 version is that the structure of the standard follows the processes within organizations. This means that the standard is easier to use. A further major advantage of the 2015 version is the reduced requirement for documented procedures. This means that the organization can develop its own individually documented QM system. The main focus of the 2015 version is on results. The most important thing is not where something is described, but if the process is effective. This encourages acceptance of the standard.

Q. What are the main differences between the 2015 and 2008 versions that the client has to consider when changing over to the more recent version?

ANS. The standard has a new structure. All clauses, from 4 to 10, with the exception of justified concessions, must be covered. Some requirements are new. These include the requirements for risk-based thinking, documented information, the context of an organization, handling of outsourced processes, stronger emphasis on management responsibility and commitment, quality control and other requirements.

Q. Why is there no longer a quality manual in 9001:2015?

ANS. System documentation continues to be required. The standard requires documented information, which also has to be controlled. However, because strict documentation of a certain kind is no longer required, the documentation can be more individually designed and adapted to the sequences and processes in the organization. The new requirements offer greater freedom for implementation and the opportunity to define processes more clearly.

Q. What is the status of the Quality Management Representative as per ISO 9001:2015?

ANS. The term "Management Representative" no longer exists as such. However, the responsibility of the management has generally increased. In addition, clear responsibility regarding processes is required.

Q. What exactly are the benefits of the standard for small enterprises/service providers?

ANS. For small enterprises/service providers, there is the possibility of focusing on processes and designing documentation individually. The term "service provider" is specifically used.

Q. Are there still Stage 1 and Stage 2 audits?

ANS. Yes, nothing has changed in that regard.

Q. What are the roles of standards?

ANS. The objective of any standard, whether it relates to the manufacture of cars, airplanes, machinery or the delivery of a service—transportation, hospitals, etc.—is the same. Standards are designed to promote, facilitate and enable consistency in a process or product; to provide assurance that the process or product output will meet requirements; to provide a uniform and predictable output every time a set of procedures are executed. Because standards assist buyers and consumers in establishing confidence levels in the products and services they procure, standards facilitate fair trade practices.

Q. What is transition period?

ANS. The International Accreditation Forum (IAF), which monitors certifications/accreditations, and the ISO Committee on Conformity Assessment (CASCO) have agreed on a three-year transition period from the publication date of ISO 9001:2015. The transition period began in September 2015 and will end in September 2018.

Q. What is the validity of ISO 9001:2008 certifications?

ANS. ISO 9001:2008 certifications will not be valid after September 2018.

Q. What are the major differences in terminology between ISO 9001:2008 and ISO 9001:2015?

ANS.

Table 22: Major differences in terminology between ISO 9001:2008 and ISO 9001:2015.

SN	ISO 9001:2008 (4th Edition)	ISO 9001:2015 (5th Edition)
1	Products	Products and Services
2	Exclusions	"Removed from the standard"
	Management Representative	Not Used (Similar responsibilities and authorities are assigned but no requirement for a single management representative)
3	Documentation, quality manual, documented procedures, records	Document Information
4	Work Environment	Environment for the Operation and Process
5	Monitoring and Measuring Equipment	Monitoring and Measuring Resources

6	Purchased Product	Externally Provided Products and Services
7	Supplier	External Provider

Q. What are the main and sub-clauses of ISO 9001:2015?

ANS.

Table 23: Main and sub-clauses of ISO 9001:2015.

Clause No.	Sub-Clause	Particulars
1	-	Scope
2	-	Normative references
3	-	Terms and definitions
4		Context of organization
	4.1	Understanding the organization and its context
	4.2	Understanding the needs and expectations of interested parties
	4.3	Determining the scope of the quality management system
	4.4	Quality management system and its processes
5	-	Leadership
	5.1	Leadership and commitment
	5.1.1	General
	5.1.2	Customer focus
	5.2	Policy
	5.2.1	Establishing the quality policy
	5.2.2	Communicating the quality policy
	5.3	Organizational roles, responsibilities and authorities
6	-	Planning
	6.1	Actions to address risks and opportunities
	6.2	Quality objectives and planning to achieve them
	6.3	Planning of changes

7	-	Support
	7.1	Resources
	7.1.1	General
	7.1.2	People
	7.1.3	Infrastructure
	7.1.4	Environment for the operation of processes
	7.1.5	Monitoring and measuring resources
	7.1.6	Organizational knowledge
	7.2	Competence
	7.3	Awareness
	7.4	Communication
	7.5	Documented information
	7.5.1	General
	7.5.2	Creating and updating
	7.5.3	Control of documented information
8	-	Operation
	8.1	Operational planning and control
	8.2	Requirements for products and services
	8.2.1	Customer communication
	8.2.2	Determining the requirements for products and services
	8.2.3	Review of the requirements for products and services
	8.2.4	Changes to the requirements for products and services
	8.3	Design and development of products and services
	8.3.1	General
	8.3.2	Design and development planning
	8.3.3	Design and development inputs
	8.3.4	Design and development controls
	8.3.5	Design and development outputs
	8.3.6	Design and development changes
	8.4	Control of externally provided processes, products and services
	8.4.1	General
	8.4.2	Type and extent of control
	8.4.3	Information for external providers

Continued

	8.5	Production and service provision
	8.5.1	Control of production and service provision
	8.5.2	Identification and traceability
	8.5.3	Property belonging to customers or external providers
	8.5.4	Preservation
	8.5.5	Post-delivery activities
	8.5.6	Control of changes
	8.6	Release of products and services
	8.7	Control of non-conforming outputs
9	-	Performance evaluation
	9.1	Monitoring, measurement, analysis and evaluation
	9.1.1	General
	9.1.2	Customer satisfaction
	9.1.3	Analysis and evaluation
	9.2	Internal audit
	9.3	Management review
	9.3.1	General
	9.3.2	Management review inputs
	9.3.3	Management review outputs
10	-	Improvement
	10.1	General
	10.2	Non-conformity and corrective action
	10.3	Continual improvement

Q. Which verbal forms have been used in QMS ISO 9001:2015?

ANS. In the QMS ISO 9001:2015, the following verbal forms are used:

1. "Shall" indicates a requirement.
2. "Should" indicates a recommendation.
3. "May" indicates a permission.
4. "Can" indicates a possibility or a capability.

Q. What would the application of the process approach in a QMS enable?

ANS. The application of the process approach in a QMS enables:

1. Understanding and consistency in meeting requirements;
2. Consideration of processes in terms of added value;
3. Achievement of effective process performance; and
4. Improvement of process approach based on evaluation of data and information.

Q. What is the PDCA Cycle?

ANS. The Plan-Do-Check-Act (PDCA) methodology can be a useful tool to define, implement and control corrective action and improvements.

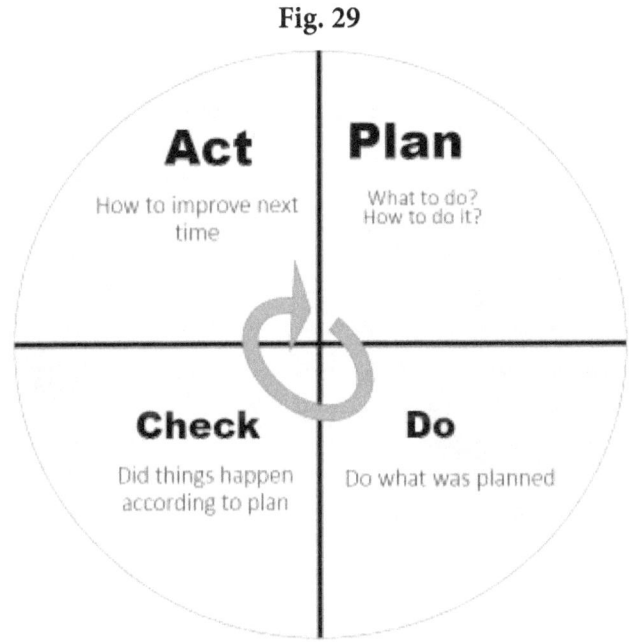

Fig. 29

Q. What is the role of the PDCA cycle in the QMS?

ANS. The PDCA cycle can be briefly described as follows:

1. Plan: establish the objectives of the system and its processes and the resources to deliver results in accordance with customer requirements and the organization's policies and identify and address risks and opportunities.
2. Do: implement what was planned.

3. Check: monitor and measure processes and the resulting products and services against policies, objectives, requirements and planned activities and report the results.
4. Act: take actions to improve performance, as necessary.

Table 24

Plan	DO	Check	Act
4.0 Context of the Organization	8.0 Operation	9.0 Performance Evaluation	10.0 Improvement
5.0 Leadership			
6.0 Planning for QMS			
7.0 Support			

Q. What are the quality objectives as per ISO 9001:2015?

ANS. The quality objectives are:

1. Be consistent with the quality policy;
2. Be measurable;
3. Take into account applicable requirements;
4. Be relevant to conformity of products and services and to enhancement of customer satisfaction;
5. Be monitored;
6. Be communicated; and
7. Be updated as appropriate.

the organization should maintain documented information on the quality objectives.

Q. How can quality objectives be achieved?

ANS. The following can be considered to achieve quality objectives:

1. What will be done?
2. What resources will be required?
3. Who will be responsible?
4. When it will be completed?
5. How will the results be evaluated?

Q. What is organizational knowledge?

ANS. It is knowledge specific to the organization, generally gained by experience. It is information that is used and shared to achieve the organization's objectives.

Organizational knowledge can be based on:

1. Internal sources—intellectual property; knowledge gained from experience; lessons learned from failures and successful projects; capturing and sharing undocumented knowledge and experience; the results of improvements in process, products and services.
2. External sources—standards, academia, conferences and gathering knowledge from customers or external providers.

Q. What are the requirements of competence in an organization as per ISO 9001:2015?

ANS. The requirements of competence in an organization as per ISO 9001:2015 are:

1. Determine the necessary competence of persons doing work under its control that affects performance and effectiveness;
2. Ensure that these persons are competent on the basis of appropriate education, training or experience; and
3. Where applicable, take action to acquire the necessary competence and evaluate the effectiveness of these actions.

Q. What QMS awareness is required for those working in the organization?

ANS. The organization must ensure that the people working under the organization's control are aware of its:

1. Quality policy;
2. Relevant quality objectives;
3. Their contribution to the effectiveness of the QMS, including the benefits of improved performance; and
4. The implications of not conforming to the QMS requirements.

Q. What communication is relevant to QMS in an organization?

ANS. The organization should determine the internal and external communications relevant to the QMS, including:

1. On what it will communicate;
2. When to communicate;
3. With whom to communicate;
4. How to communicate; and
5. Who communicates.

Q. Which activities must be addressed during the control of documented information?

ANS. For the control of documented information, the following activities must be addressed:

1. Distribution, access, retrieval and use;
2. Storage and preservation, including preservation of legibility;
3. Control of change (for example, version control); and
4. Retention and disposition.

Documented information retained as evidence of conformity must be protected from unintended alterations.

Q. What are the requirements for communication on products and services?

ANS. As per the new QMS, communication with customers should include:

1. Providing information relating to products and services;
2. Handling enquiries, contracts or orders, including changes;
3. Obtaining customer feedback relating to products and services, including customer complaints;
4. Handling or controlling customer property; and
5. Establishing specific requirements for contingency actions, when relevant.

Q. What is preservation as per QMS?

ANS. Preservation can include identification, handling, contamination control, packaging, storage, transmission or transportation and protection.

Q. What are post-delivery activities as per the new QMS?

ANS. The organization should meet requirements for post-delivery activities associated with products and services. In determining the extent of post-delivery activities that are required, the organization should consider:

1. Statutory and regulatory requirements;
2. The potential undesired consequences associated with its products and services;
3. The nature, use and intended lifetime of its products and services;
4. Customer requirements; and
5. Customer feedback.

Q. What are the inputs for management review as per ISO 9001:2015?

ANS. The management review should be planned and carried out taking into consideration:

1. Status of actions from previous management reviews;
2. Changes in external and internal issues that are relevant to the QMS;
3. Information on the performance and effectiveness of the QMS, including trends in
    a. Customer satisfaction and feedback from relevant interested parties,
    b. The extent to which quality objectives have been met,
    c. Process performance and conformity of products and services,
    d. Non-conformities and corrective actions,
    e. Monitoring and measurement results,
    f. Audit results, and
    g. Performance of external providers;
4. The adequacy of resources;

5. Effectiveness of actions taken to address risks and opportunities; and
6. Opportunities for improvement.

Q. What are the reference ISO standards for QMS?
ANS. The references for QMS are:
1. ISO 9000 QMS—fundamental and vocabulary.
2. ISO 9001 QMS—specifies requirements.
3. ISO 9004 managing for the sustained success of an organization—a quality management approach.
4. ISO 10001 quality management—customer satisfaction: guidelines for codes of conduct for organizations.
5. ISO 10002 quality management—customer satisfaction: guidelines for complaints handling in organizations.
6. ISO 10003 quality management—customer satisfaction: guidelines for dispute resolution external to organizations.
7. ISO 10004 quality management—customer satisfaction: guidelines for monitoring and measuring.
8. ISO 10005 QMS—guidelines for quality plans.
9. ISO 10006 QMS—guidelines for quality management in projects.
10. ISO 10007 QMS—guidelines for configuration management.
11. ISO 10008 quality management—customer satisfaction: guidelines for business to consumer electronic commerce transactions.
12. ISO 10012 measurement management systems—requirements for measurement processes and measuring equipment.
13. ISO/TR 10013—guidelines for quality management system documentation.
14. ISO 10014 quality management—guidelines for realizing financial and economic benefits.
15. ISO 10015 quality management—guidelines for training.

16. ISO/TR 10017—guidance on statistical techniques for ISO 9001:2000.
17. ISO 10018 quality management—guidelines on people involvement and competence.
18. ISO 10019—guidelines for the selection of QMS consultant and use of their services.
19. ISO 19011—guidelines for auditing management systems.

Q. What are the key factors that do not need to be changed under the new QMS?

ANS. Organizations do not need to:

a. REMOVE their management representatives. While there is no requirement in ISO 9001:2015 for a management representative, this does not prevent the organization from choosing to retain this role if they so wish. However, some of the duties (responsibilities) traditionally assigned to the management representative by top management will, in future, need to be undertaken directly by top management themselves.

b. RELEGATE their quality manuals and documented procedures to the dustbin. ISO 9001:2015 has no requirement for the organization to have and use either a quality manual or documented procedures. However, if this documentation is in place, needed and working well, there is no need for it to be withdrawn.

c. RENUMBER or rename existing QMS documentation to correspond to the new clause references. Although an organization may choose to carry out a renumbering/renaming exercise, it is down to them to determine whether the benefits gained from renumbering/renaming will exceed the effort involved in the change.

d. RESTRUCTURE their management systems to follow the sequence of requirements as set out in the standard. Providing all of the requirements contained in the standard are met, the organization's system will be compliant.

e. REFRESH existing documentation to use the new terms and definitions contained within ISO 9000:2015. Once again, the organization is free to make the judgement as to whether this effort would be worthwhile. If the organization is more comfortable using their own terminology, for instance "records" instead of "documented information," or "supplier" rather than "external provider," then this is perfectly acceptable.

Q. Who are interested parties and what are their needs and expectations as per the new QMS (ISO 9001:2015)?

ANS.

Table 25: Interested parties and their needs and expectations as per the new QMS (ISO 9001:2015).

Interested Parties	Needs and Expectations
Customers	Quality, price and delivery of products and services
Owners/Shareholders	Sustained profitability
	Transparency
People in the organization	Good work environment
	Job security
	Recognition and reward
Suppliers and partners	Mutual benefit and continuity
Society	Environmental protection
	Ethical behavior
	Compliance with statutory and regulatory requirements

Q. What is risk?

ANS. According to ISO 31000, *risk* is the "effect of uncertainty on objectives" and an *effect* is a positive or negative deviation from what is expected. Risk is often expressed in terms of a combination of the consequences of an event and the associated likelihood of occurrence.

Q. What is risk management?

ANS. *Risk management* refers to a coordinated set of activities and methods used to direct an organization and to control the many risks that can affect its ability to achieve objectives.

Q. What is risk framework?

ANS. According to ISO 31000, a risk management framework is a set of components that support and sustain risk management throughout an organization.

There are two types of components:

1. Foundational arrangements (include risk management policy, objectives, mandate and commitment) and
2. Organizational arrangements (include plans, relationships, accountabilities, resources, processes and activities).

Q. What is risk-management policy?

ANS. A policy statement defines a general commitment, direction or intention. A risk-management policy statement expresses an organization's commitment to risk management and clarifies its general direction or intention.

Q. What is risk attitude?

ANS. An organization's risk attitude defines its general approach to risk. It (and its risk criteria) influence how risks are assessed and addressed. This influences whether or not risks are taken, tolerated, retained, shared, reduced or avoided and whether or not risk treatments are implemented or postponed.

Q. Who is a risk owner?

ANS. A risk owner is a person or entity given the authority to manage a particular risk and accountable for doing so.

Q. What is a risk-management plan?

ANS. A risk-management plan is a high-level document that sets out the main parameters of how risks are to be managed. It includes the description of the project, the types of risks to be considered, the risk process, including the tools and techniques to be followed, the risk reporting procedure and the major risks envisaged.

Q. What is risk-management process?

ANS. According to ISO 31000, a risk-management process is one that systematically applies management policies, procedures and practices to a set of activities intended to establish the context; communicate and consult with stakeholders; and identify, analyze, evaluate, treat, monitor and review risk.

The risk-management process consists of five major stages:

1) risk awareness,
2) risk identification,
3) risk assessment,
4) risk evaluation, and
5) risk management.

Q. What are the steps for risk assessment?

Ans. The steps for risk assessment are:

1. Risk planning,
2. Risk identification,
3. Risk evaluation,
4. Risk response, and
5. Risk monitoring and control.

Q. What is risk identification?

ANS. Risk identification involves finding, recognizing and describing the risks that could affect the achievement of an organization's objectives. It is used to identify possible sources of risk in addition to the events and circumstances that could affect the achievement of objectives. It also includes the identification of possible causes and potential consequences.

Q. What is risk source?

ANS. A risk source has the intrinsic potential to give rise to risk. A risk source is where a risk originates. Potential sources of risk include at least the following: commercial relationships and obligations, legal expectations and liabilities, economic shifts and circumstances, technological innovations and upheavals, political changes and trends, natural events and forces, human frailties and tendencies and management shortcomings and excesses. All of these elements could potentially generate a risk that must be managed.

Q. What is consequence?

ANS. A consequence is the outcome of an event and has an effect on objectives. A single event can generate a range of consequences which can have both positive and negative effects on objectives. Initial consequences can also escalate through knock-on effects.

Q. What is likelihood in risk assessment?

ANS. In risk management terminology, likelihood is used to refer to the chance of something happening, whether defined, measured or determined objectively or subjectively, qualitatively or quantitatively and described using general terms or mathematically (such as probability or frequency over a given time period).

Q. What is risk profile?

ANS. A risk profile is a written description of a set of risks. A risk profile can include risks that the entire organization must manage or only those that a particular function or part of the organization must address.

Q. What is risk analysis?

ANS. Risk analysis is a process used to understand the nature, sources and causes of the risks identified and to estimate the level of risk. It is also used to study impacts and consequences and to examine the controls that currently exist. Risk analysis provides the basis for risk evaluation and decisions about level of risk. Risk analysis includes risk estimation.

Q. What is risk criteria?

ANS. Risk criteria are terms of reference and are used to evaluate the significance or importance of an organization's risks. They are used to determine whether a specified level of risk is acceptable or tolerable. Risk criteria should reflect the organization's values, policies and objectives; should be based on its external and internal context; should consider the views of stakeholders and should be derived from standards, laws, policies, and other requirements.

Q. What is level of risk?

ANS. The level of risk is its magnitude. It is estimated by considering and combining consequences and likelihoods. A level of risk can be assigned to a single risk or to a combination of risks. A consequence is the outcome of an event and has an effect on objectives. Likelihood is the chance that something might happen.

Q. What is risk evaluation?

ANS. Risk evaluation is a process used to compare risk analysis results with risk criteria in order to determine whether or not a specified level of risk is acceptable or tolerable.

Q. What is risk treatment?

ANS. Risk treatment is a risk-modification process. It involves selecting and implementing one or more treatment options. Once a treatment has been implemented, it becomes a control or it modifies existing controls. There are many treatment options: avoiding the risk, reducing the risk, removing the source of the risk, modifying the consequences, changing the probabilities, sharing the risk with others, retaining the risk or even increasing the risk to pursue an opportunity. Risk treatments can create new risks or modify existing ones.

Q. What is risk control?

ANS. A control is any measure or action that modifies risk. Controls include any policy, procedure, practice, process, technology, technique, method or device that modifies or manages risk. Risk treatments become controls or modify existing controls, once they have been implemented.

Q. What is residual risk?

ANS. It is the risk left over after a risk-treatment option has been implemented. It is the risk remaining after reducing risk, removing its source, modifying consequences, changing probabilities, transferring risk or retaining risk. Residual risk is also known as "retained risk."

Q. What is monitoring?

ANS. To monitor means to supervise and to continually check and critically observe. It means to determine the current status and to assess whether or not required or expected performance levels are actually being achieved. Monitoring can be applied to a risk-management framework, risk-management process, risk or control.

Q. What is review of risk?

ANS. A review is an activity. Reviews are carried out to determine whether something is a suitable, adequate and effective way of achieving established objectives. Risk review examines risk-management policies and plans as well as risks, risk criteria, risk treatments, controls, residual risks and the risk-assessment process.

Q. What are the principles of risk management?

ANS. The principles of risk management are that it

1. Creates and protects value;
2. Is an integral part of all organizational processes;
3. Is part of decision making;
4. Explicitly addresses uncertainty;
5. Is systematic, structured and timely;
6. Is based on the best available information;
7. Is tailored;
8. Takes human and cultural factors into account;
9. Is transparent and inclusive;
10. Is dynamic, iterative and responsive to change; and
11. Facilitates continual improvement of the organization.

Q. What should be addressed in the risk assessment?

ANS. The following basic questions should be addressed in the risk assessment:

1. What might go wrong?
2. What is the nature of possible risks?
3. What is the probability of their occurrence and how easy is it to detect them?

4. What are the consequences (the severity)?

Q. Can risk-management framework be represented diagrammatically?
ANS. Yes.

Fig.30: Risk-management framework.

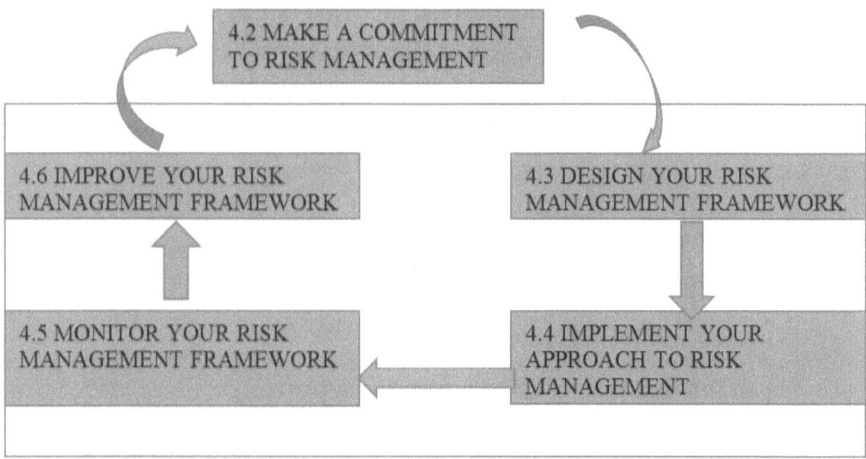

Q. Can the risk-management process be represented diagrammatically?
ANS. Yes.

Fig. 31: Risk-management process.

Q. What is the risk matrix?

ANS. The risk-rating table or risk matrix is show below:

Table 26: Risk-rating based on likelihood and consequence.

Likelihood	Numbers	Consequences				
		Insignificant	Minor	Moderate	Major	Extreme
		1	2	3	4	5
Almost Certain (Frequent)	5	5	10	15	20	25
Likely (Probable)	4	4	8	12	16	20
Possible (Occasional)	3	3	6	9	12	15
Unlikely (Uncommon)	2	2	4	6	8	10
Rare (Remote)	1	1	2	3	4	5

Table 27: Risk rating—management action required.

Risk Score	Risk Rating	Action Required
1–4	Low Risk	Manage by routine procedures; report to local managers; monitor and review locally as necessary.
5–10	Medium Risk	Assess the risk; determine whether current controls are adequate or if further action or treatment is needed; monitor and review locally, for example, through regular business practices or local area meetings.
11–15	High Risk	Risk to be given appropriate attention and demonstrably managed, reported to vice-chancellor or other senior executives/management committees as necessary.
16–25	Extreme Risk	Immediate attention and response needed; requires a risk-assessment and management plan prepared by relevant senior managers for vice-chancellor; risk oversight by council or nominated standing committee or management committee.

Q. What are some risk-management tools?

ANS.

1. Diagram analysis.
2. Risk ranking and filtering.
3. Fault tree analysis.
4. Hazard operability analysis.
5. Hazard Analysis Critical Control Point (HACCP).
6. Failure Mode Effects Analysis (FMEA).

Q. Can quality risk management process be represented diagrammatically?

ANS. Yes.

Fig. 32: Quality risk-management process.

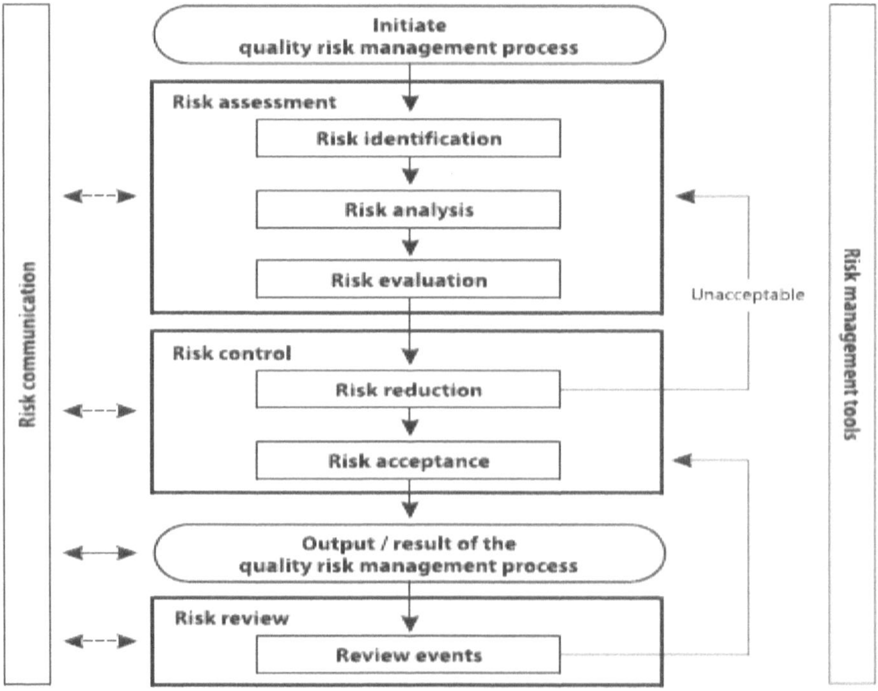

Q. What is a risk register?

ANS.

Table 28: An example of a risk register.

S N	Process/ Activity	Risk Description	Inherent Rating			Risk Treatments in Place (to eliminate or minimize risk)	Residual Rating (including controls)		
			Severity	Likelihood	Risk		Severity	Likelihood	Risk

Q. What is a quality plan register?

ANS.

Table 29: An example of a quality plan register:

SN	Product Category	Product	Test	Para meters	Externally/ Internally	Frequency	Responsibility

FOOD SAFETY MANAGEMENT SYSTEM

This is a basic idea about food safety management system, HACCP system, Principle of HACCP, requirement of FSMS, clauses and CCP, OPRP and PRP requirement etc.

Q. What is HACCP?

ANS. Hazard Analysis Critical Control Point (HACCP) is a process-control system designed to identify and prevent microbial and other hazards in food production.

Another definition is "**HACCP** is a management system in which food safety is addressed through the analysis and control of biological, chemical, and physical hazards from raw material production, procurement and handling, to manufacturing, distribution and consumption of the finished product." (HACCP Manual)

Developing, implementing and maintaining a **HACCP** system is the industry's responsibility. This is because food manufacturers have the most control over the products they manufacture, so they have the greatest impact on the safety of their products.

Q. Why is HACCP important?

ANS. HACCP is important because it prioritizes and controls potential hazards in food production. By controlling major food risks, such as microbiological, chemical and physical contaminants, the industry can better assure consumers that its products are as safe as good science and technology allows. By reducing foodborne hazards, public health protection is strengthened.

Q. What are the major food hazards?

ANS. Many public opinion studies report that consumers are concerned primarily about chemical residues, such as from pesticides and antibiotics.

These hazards are nearly non-existent. The more significant hazards facing the food industry today are microbiological contaminants, such as *Salmonella, E. coli O157:H7, Listeria, Campylobacter* and *Clostridium botulinum*. HACCP is designed to focus on and control the most significant hazards.

Q. Is HACCP new?

ANS. HACCP is not new. It was first used in the 1960s by the Pillsbury Company to produce the safest and highest quality food possible for astronauts in the space program. The National Academy of Sciences, National Advisory Committee for Microbiological Criteria for Foods, and the Codex Alimentarius have endorsed HACCP as the best process control system available today.

Q. What are the seven principles of HACCP?

ANS. The seven principles of HACCP are:
1. Conduct a hazard analysis;
2. Identify critical control points;
3. Establish critical limits;
4. Establish monitoring;
5. Establish corrective actions;
6. Establish documentation and records; and
7. Establish verification.

Q. What is a critical control point in food safety?

ANS. A critical control point (CCP) is a point, step or procedure in a food manufacturing process at which control can be applied and, as a result, a food safety hazard can be prevented, eliminated or reduced to an acceptable level.

Q. What is aspect?

ANS. It is an element of the food business operation (products, processes, PRP, services, etc.) that can interact with the food safety.

Q. What is control?

ANS. It means to take all necessary actions to ensure and maintain compliance with criteria established in the HACCP plan. It refers to the state wherein correct procedures are being followed and criteria are being met.

Q. What is control measure?

ANS. Any action and activity that can be used to prevent or eliminate a food safety hazard or reduce it to an acceptable level.

Q. What is corrective action?

ANS. Any action to be taken when the results of monitoring at the CCP indicate a loss of control.

Q. What are critical limits?

ANS. It is a criterion that separates acceptability from non-acceptability. A maximum and/or minimum value to which a biological, chemical or physical parameter must be controlled at a CCP to prevent, eliminate or reduce to an acceptable level the occurrence of a food safety hazard.

Note: This criterion defines the limiting values for the product or process parameter(s) under consideration for monitoring.

Q. What is a flow diagram?

ANS. A systematic representation of the sequence of steps or operations used in the preparation, processing, manufacturing, packaging, storage, transportation, distribution, handling or offering for sale of a particular food item.

Q. Who is a food handler?

ANS. Any person who directly handles packaged or unpacked food, food equipment and utensils or food contact surfaces and is, therefore, expected to comply with food hygiene requirements.

Q. What is food hygiene?

ANS. All conditions and measures necessary to ensure the safety and suitability of food at all stages of the food chain.

Q. What is food safety?

ANS. Assurance that food will not cause harm to the consumer when it is prepared and/or eaten according to its intended use. Food safety is about making sure that food products are safe to eat.

Q. What is an HACCP audit?

ANS. A systematic and independent examination to determine whether the HACCP system, including the HACCP plan and related results, comply with planned arrangements, are implemented effectively and are suitable for the achievement of stated objectives.

Note: Examination of hazard analysis is an essential element of the HACCP audit.

Q. What is an HACCP team?

ANS. It is made up of the group of people responsible for developing, implementing and maintaining the HACCP system.

Q. What is an HACCP plan?

ANS. It is a document prepared in accordance with the principles of HACCP to ensure control of hazards significant for food safety in the segment of the food chain under consideration. It is the written document based upon HACCP principles that delineates the procedures to be followed to assure control of specific process or procedure.

Q. What is hazard?

ANS. It refers to a biological, chemical or physical agent in, or condition of, food with the potential to cause an adverse health effect. A biological, chemical or physical agent that is reasonably likely to cause a food to be unsafe for consumption.

Q. What is hazard analysis?

ANS. It is the process of collecting and evaluating information on hazards and conditions leading to their presence, to decide which are significant for food safety and should, therefore, be addressed in the HACCP plan.

Q. What is monitoring?

ANS. The act of conducting a planned sequence of observations or measurement of control parameters to assess whether a CCP is under control is called monitoring.

Q. What is the Pre-Requisite Program (PRP)?

ANS. It refers to any specified and documented activity or facility implemented in accordance with the Codex General Principles of food hygiene, good manufacturing practice and appropriate food legislation to establish basic conditions that are suitable for the production and handling of safe food at all stages of the food chain. It involves procedures, including Good Manufacturing Practices that address operational conditions providing the foundation for the HACCP system.

Q. What is preventive action?

ANS. Any measure or activity used to prevent, eliminate or reduce the recurrence of causes for existing deviations, defects or any other undesired situation with respect to food safety is called preventive action.

Q. What is food risk?

ANS. It refers to the probability of causing an adverse health effect by the occurrence and severity of a particular hazard in food when prepared and consumed according to its intended use.

Q. What is target value?

ANS. It is the value of the product or process parameter(s) to be monitored, targeted within action-limit values (the range of acceptable variations) and certainly within critical limit values, thus securing a safe product.

Q. What is a Step?

ANS. It is a point, procedure, operation or stage in the food chain, including raw materials, from primary production to final consumption.

Q. What is validation?

ANS. Obtaining evidence (in advance) that the specific and general control measures of the HACCP plan are effective is known as validation.

Q. What is verification?

ANS. It refers to the application of methods, procedures, tests and other evaluations, in addition to monitoring, to determine compliance with the specifications laid down in the HACCP plan and the effectiveness of the HACCP-based food safety system.

Q. What are the pre-requisites for HACCP?

ANS. The pre-requisites for HACCP are:

1. Good Manufacturing Practices (GMP)
2. Good Hygiene Practices (GHP)
3. Good Sanitary Practices (GSP)
4. Good Laboratory Practices (GLP)
5. Good Documentation Practices (GDP)

Q. Who is at risk?

ANS. The following are at risk:

1. Young children;
2. Elderly women;
3. Pregnant women (and fetuses);
4. Patients taking antibiotics and with low immunity due to HIV/AIDS, cancer treatment, organ transplant, etc.; and
5. Persons living in institutional settings such as hospitals and nursing homes.

Q. What are the benefits of HACCP?

ANS. HACCP

1. Is preventive rather than remedial;
2. Offers control through easily monitored features;
3. Provides a more rapid response;
4. Reduces final product testing; and
5. Identifies potential hazards.

Q. What is the responsibility of the food safety team?

ANS. The food safety team is responsible for

1. Creating awareness on the food safety management system among employees;
2. Performing hazard analysis and preparing HACCP plans;

3. Planning and performing verification and validation activities;
4. Planning and performing internal quality audits;
5. Ensuring compliance with statutory and regulatory requirements;
6. Identifying resources required for implementation and management of the food safety management system.
7. Ensuring strict implementation of the organization's hygiene practices; and
8. Initiating corrective and preventive action on product rejections, customer complaints and recalls.

Q. What is the responsibility of the leader of the food safety team?

ANS. The leader of the food safety team is responsible for:

1. Managing a food safety team and organizing its work;
2. Ensuring relevant training and education of the food safety team members;
3. Ensuring that the food safety management system is established, implemented, maintained and updated;
4. Reporting to the organization's top management on the effectiveness and suitability of the food safety management system; and
5. Reviewing and implementing changes to the food safety management system.

Q. What is GMP?

ANS. It refers to all features of design, construction of facility, utility and equipment, the maintenance of the same, cleaning and sanitation in a food plant that facilitate the production of safe and wholesome foodstuff.

Q. What is the 12-step approach of HACCP (as per Codex Alimentarius)?

ANS. It involves the following steps:

1. Assemble HACCP team.
2. Describe product.
3. Identify intended use.

4. Construct a flow diagram.
5. Verify the diagram through on-site confirmation.
6. List out potential hazards associated with each step and preventive measures (apply HACCP decision tree).
7. Determine CCP (establish target tolerances).
8. Establish critical limits for each CCP.
9. Establish a monitoring system for each CCP.
10. Establish a system for corrective action.
11. Establish documentation and record keeping.
12. Establish verification procedures.

Q. What is PRP?

ANS. Some examples of pre-requisites are:
1. Construction and layout of the buildings;
2. Layout of the premises and workplace;
3. Utilities—air, water and energy;
4. Waste disposal facilities;
5. Suitable equipment and facilities for their cleaning and maintenance;
6. Management systems for purchased materials;
7. Measures for prevention of cross contamination;
8. Facilities for cleaning and sanitizing;
9. Facilities for pest control;
10. Facilities to ensure personnel hygiene;
11. Rework;
12. Procedures for product recall;
13. Facilities for warehousing;
14. Systems for providing product information/consumer awareness; and
15. Systems for food defense, bio vigilance and bio terrorism.

Q. What are the clauses and sub clauses of ISO 22000: 2005?
ANS.

Table 30: Clauses and sub clauses of ISO 22000: 2005.

Clause No.	Sub-Clause	Particulars
	-	Introduction
1	-	Scope
2	-	Normative references
3	-	Terms and definitions
4	-	Food safety management system
	4.1	General requirements
	4.2	Document requirements
	4.2.1	General
	4.2.2	Control of documents
	4.2.3	Control of records
5	-	Management responsibility
	5.1	Management commitment
	5.2	Food safety policy
	5.3	Food safety management system planning
	5.4	Responsibility and authority
	5.5	Food safety team leader
	5.6	Communication
	5.6.1	External communication
	5.6.2	Internal communication
	5.7	Emergency preparedness and response
	5.8	Management review
	5.8.1	General
	5.8.2	Management review input
	5.8.3	Management review output
6	-	Resource management
	6.1	Provision of resources
	6.2	Human resources

	6.2.1	General
	6.2.2	Competence, awareness and training
	6.3	Infrastructure
	6.4	Work environment
7	-	Planning and realization of safe products
	7.1	General
	7.2	Prerequisite program
	7.3	Preliminary steps to enable hazard analysis
	7.3.1	General
	7.3.2	Food safety team
	7.3.3	Product characteristics
	7.3.4	Intended use
	7.3.5	Flow diagrams, process steps and control measures
	7.4	Hazard analysis
	7.4.1	General
	7.4.2	Hazard identification and determination of acceptable levels
	7.4.3	Hazard assessment
	7.4.4	Selection and assessment of control measures
	7.5	Establishing the operational PRPs
	7.6	Establishing the HACCP plan
	7.6.1	HACCP plan
	7.6.2	Identification of CCPs
	7.6.3	Determination of critical limits of CCPs
	7.6.4	System for the monitoring of CCPs
	7.6.5	Actions when monitoring results exceed critical limits
	7.7	Updating of preliminary information and documents specifying the PRPs and HACCP plan
	7.8	Verification planning
	7.9	Traceability system

Continued

	7.10	Control of non-conformity
	7.10.1	Corrections
	7.10.2	Corrective actions
	7.10.3	Handling of potential unsafe products
	7.10.4	Withdrawals
8	-	Validation, verification and improvement of the food safety management system
	8.1	General
	8.2	Validation of control measure combinations
	8.3	Control of monitoring and measuring
	8.4	Food safety management system verification
	8.4.1	Internal audit
	8.4.2	Evaluation of individual verification results
	8.4.3	Analysis of results of verification activities
	8.5	Improvement
	8.5.1	Continual improvement
	8.5.2	Updating the food safety management system

Q. What is safe food?

ANS. Food that is free of contaminants and will not cause illness or harm is referred to as safe food.

Q. What are the benefits of good food hygiene?

ANS. Good food hygiene has two main benefits:

1. Compliance with the law and
2. Fewer customer complaints.

food poisoning is an illness caused by eating contaminated food.

Q. What are the cross references between HACCP principles and application steps and clauses of ISO 22000?

ANS.

Table 31

HACCP Principles	HACCP Application Steps		ISO 22000 :2005	
	Assemble HACCP team	Step 1	7.3.2	Food safety team
	Describe product	Step 2	7.3.3 7.3.5.2	Product characteristics Description of process steps and control measures
	Identify intended use	Step 3	7.3.4	Intended use
	Construct flow diagram	Step 4	7.3.5.1	Flow diagrams
	On-site confirmation of flow diagram	Step 5		
Principle 1 Conduct a hazard analysis	List all potential hazards Conduct a hazard analysis Consider control measures	Step 6	7.4 7.4.2 7.4.3 7.4.4	Hazard analysis Hazard identification and determination levels Hazard assessment Selection and assessment of control measures
Principle 2 Determine CCPs	Determine CCPs	Step 7	7.6.2	Identification of CCPs
Principle 3 Establish critical limits	Establish critical limits for each CCP	Step 8	7.6.3	Determination of critical limits for CCPs

Continued

Principle 4 Establish a system to monitor control of the CCP	Establish a monitoring system for each CCP	Step 9	7.6.4	System for the monitoring of CCPs
Principle 5 Establish the corrective action to be taken when monitoring indicates that a particular CCP is not under control	Establish corrective actions	Step 10	7.6.5	Actions when monitoring results exceed critical limits
Principle 6 Establish procedures for verification to confirm that the HACCP system is working effectively	Establish verification procedures	Step 11	7.8	Verification planning
Principle 7 Establish documentation concerning all procedures and records appropriate to these principles and their application	Establish documentation and record keeping	Step 12	4.2 7.7	Documentation requirements Updating of preliminary information and documents specifying the PRPs and the HACCP plan

Q. How can good food hygiene be ensured?

ANS. To ensure good food hygiene use the following steps:

1. Protect food from contamination;
2. Destroy bacteria;
3. Discard unfit or contaminated food; and
4. Prevent multiplication of bacteria.

Q. What is incubation period/onset time?

ANS. It refers to the period between eating contaminated food and the appearance of first symptoms.

Q. Where should high risk food be stored?

ANS. Such food should be stored in a fridge.

Q. What are the general principles of GHP?

ANS. The general principles of GHP are:

1. Primary production.
2. Establishment, design and facilities.
3. Control of operations.
4. Maintenance and sanitation.
5. Personal hygiene.
6. Transportation.
7. Product information and consumer awareness.
8. Training.

Q. Why should diamond jewelry not be worn in food premises?

ANS. Stones and small pieces of metal may end up on food. They harbor dirt and bacteria.

Q. Why should smoking not be allowed in a food room?

ANS. Smoking encourages coughing, which may result in bacteria being transferred from the mouth to the food.

Q. When selecting a site for a new business, what should be considered in relation to food safety?

ANS. When selecting a site for a business, the availability of electricity and gas, access to waste removal and food deliveries are critical from a food safety perspective.

Q. Why is physical control of pests preferable to chemical control?

ANS. With physical control there is no risk of contaminating the food with pesticides; the pest is caught, either dead or alive.

Q. What is the correct order to put on protective clothing?

ANS. Precautions against contamination require that the following order be maintained when putting on protective clothing:

1. First, hat;
2. Second, coat; and
3. Third, shoes

Q. For how long can food on display be below 63 °C?

ANS. It can be left below 63 °C for two hours.

Q. What is food-borne illness?

ANS. Food-borne illness is a disease carried or transmitted to people by food.

Q. What is food-borne illness outbreak?

ANS. It is when two or more people experience the same illness after eating the same food.

Q. In how many ways can food be contaminated?

ANS. Food can be contaminated in four ways:

1. Time-temperature control: TCS foods are left in the danger zone for more than four hours.
2. Cross contamination: contaminants cross to a food that is not going to be cooked any further.

3. Poor personal hygiene: food handlers cause food to be contaminated.
4. Poor cleaning and sanitization.

Q. What is ALERT?

ANS. ALERT is an FDA Defense Tool. It stands for

Assure:	make sure products received are from safe sources;
Look:	monitor the security of products in the facility;
Employees:	know who is in your facility;
Reports:	keep information related to food defense accessible; and
Threat:	develop a plan for responding to suspicious activity or a threat to the operation.

Q. What are food allergies?

ANS. An allergic reaction to food could include itching, tightening of the throat, wheezing, hives, swelling, diarrhea, vomiting, cramps and loss of consciousness or even death. Managers and employees should be aware of the most common food allergens: milk; eggs; fish; shellfish including lobster, shrimp and crab; wheat; soy; peanuts and tree nuts such as almonds, walnuts and pecans.

Q. Can HACCP guidelines and plans be used by the food industry?

ANS. There are seven HACCP principles that must be followed to implement HACCP. Every food production process in a plant will need an individual HACCP plan that directly impacts the specifics of the product and process. Government and industry groups are developing some generic HACCP models that provide guidelines and directions for developing plant-, process- and product-specific HACCP systems.

Q. How does HACCP differ from traditional food safety systems?

ANS. HACCP places greater responsibility on the food producer to identify and control hazards and document the effectiveness of the system. In addition, it requires constant verification that the system is working.

Q. What are the advantages of adopting the HACCP system?

ANS. The primary purpose of a HACCP system is to protect people from food-borne illnesses, but the benefits of the system also extend to the company:

1. Increased confidence in products;
2. Ability to reach markets and customers that require a HACCP-based system;
3. Reduced liability;
4. Effective process management; and
5. Improved quality and consistency.

   Safety is enhanced by a proactive approach of continuous monitoring of food safety controls and documentation of results and corrective actions. This monitoring takes place in "real time" rather than a reactive, after-the-fact approach.

Q. How does HACCP enhance food safety?

ANS. HACCP requires monitoring to reveal when food safety limits have been violated. This results in taking corrective action to reinstate control and documented procedures to prevent recurrence. The operation of the system is constantly verified.

Q. What food safety issues does HACCP address?

ANS. HACCP evaluates and addresses potential biological, physical and chemical hazards. These may be introduced from raw materials, the process, equipment, the environment and employees.

Q. Could the critical control limit be a range?

ANS. No, the critical limit must have a specific cut-off.

Q. What is quality safe food?

ANS. Safe food is free from visible contamination or spoilage. Quality safe food must also be free from "invisible" contamination by microbes such as bacteria and mold, which may cause illness or produce dangerous spores or toxins.

Q. How does microbial food contamination occur?

ANS. Microbial food contamination occurs through:

1. Poor food-processing systems before the food items are received into store;
2. Contact with unclean surfaces in food-preparation areas (floors, benches, slicers and knives);
3. Contact with domestic animals, insects or rodents, cockroaches, flies or rats;
4. Cooks and other food handlers who neglect to wash their hands and who have a low standard of personal hygiene; and
5. Raw food coming into contact with cooked food (raw meat and unwashed vegetables or fruits).

Q. In which type of environment does bacteria grow well?

ANS. Bacteria grow well in

1. Suitable foods (sugars, alcohols and amino acids and foods that are low in carbohydrates and fats);
2. Sufficient moisture (bacteria mostly require higher water activity levels for growth than molds);
3. Favorable temperatures (body temperature is ideal);
4. Time (if the above conditions are right, the number of bacteria present in food can double approximately every twenty minutes); and
5. Near-neutral pH (acid/alkali conditions).

Q. What are the three types of contamination that can occur?

ANS. The three type of contamination that can occur are:

1. Physical contamination—from food handlers (for example, from jewelry and hair), cleaning implements (steel wool, scourers and cloths), premises (dust, flaking paint), faulty equipment (nuts, bolts, screws and filings), insects and vermin (dead or live insects, rodent droppings) and from the food itself (seeds or stones that may be present in the raw foods).

2. Chemical contamination—from poor cleaning practices (for example, incorrectly diluted chemicals), incorrectly stored chemicals (storing chemicals in food containers), food handlers (perfumes) and the use of inappropriate chemicals in the premises and equipment (diesel-powered forklifts in a store area, nonfood-grade lubricants in equipment).
3. Biological contamination—harmful bacteria (pathogens), viruses, parasites and molds.

Q. What is HACCP not?

ANS. HACCP is not:

1. A quality control system or
2. A government program.

Q. Who is responsible for the verification of flow diagrams?

ANS. The food safety team assigns responsibility for verifying the accuracy of the flow diagrams and the responsible person signs the flow diagram as an indication of successful validation.

Q. What are some examples of critical limits?

ANS. Some examples of critical limits are pH, temperature, time and humidity.

Q. What records are required for HACCP?

ANS. The following records are required for HACCP:

1. HACCP plan;
2. hazard analysis;
3. monitoring records and data;
4. testing data and results;
5. records of corrective action;
6. non-conforming product disposal records;
7. validation records; and
8. audit reports.

Q. What are the general principles of food hygiene?

ANS. The general principles of food hygiene are:

1. primary production;
2. design and facility;
3. control of operation;
4. maintenance and sanitation;
5. personal hygiene;
6. transportation;
7. product information and consumer awareness; and
8. training.

Q. What is the history of HACCP?

ANS. The history of HACCP is as follows:

1. It was first used in the US space program to ensure food safety for astronauts without relying on end-product testing (early 1970s).
2. HACCP was adopted in 1973 by the USFDA for low-acid canned food regulations (pH > 4.6).
3. The USFDA made HACCP mandatory for all seafood processors in the US as well as for foreign plants exporting to the US (1997).
4. The Canadian Food Inspection Agency (CFIA) made HACCP mandatory for all Canadian seafood processors (1998).
5. CFIA also implemented mandatory HACCP for the meat and poultry industry (2007).

Q. What is the difference between auditable standards and guidelines?

ANS. A standard specification against which an independent audit can be conducted is known as an auditable standard whereas guidelines are best practices available for selection for achieving a certain objective.

Q. What is the correlation between various food standards?

ANS.

Table 32: Correlation between various food standards.

SN	Standard	Application
1	ISO 22001:2007	Guidance for application of ISO 9001:2000 in food and drink industry (previously ISO 15161:2002).
2	ISO 22003:2007	Food safety management systems—requirements for bodies providing audit and certification of food safety management systems.
3	ISO 22004:2006	Guidance on application for ISO 22000:2005.
4	ISO 22005:2007	Guidance on traceability and transparency in the food chain.
5	ISO/DIS 22006	Quality management system—guidelines on the application of ISO 9001:2000 for crop production.
6	ISO/WD 22008	Food irradiation—good processing practices for irradiation of foods intended for human consumption.

Q. What are some measures that can be taken to ensure personal hygiene standards are met?

ANS. Employees must ensure that they:

- Shower/bath and change clothing daily.
- Shampoo their hair regularly, ensure their hair brush is kept clean and that they wear clean head coverings (especially if their hair is long).
- Keep hands and fingernails clean and ensure that any sores or cuts are disinfected, cleaned and covered with a waterproof dressing and/or wear surgical gloves when handling food.
- Keep fingernails short and clean.
- Wear clean, well-fitting shoes for work (they must not wear their work shoes unless they are at work).
- Their teeth are clean and maintained.
- Wash their hands thoroughly with soap and hot water and dry their hands with a paper towel or air dryer (old linen cloths that are lying around must not be used; they might negate any effect washing has had).

- Wash their hands regularly. They should wash their hands:
  - before starting work,
  - after a work break,
  - after going to the toilet,
  - after handling anything dirty or unwashed,
  - after handling garbage or chemicals,
  - after handling uncooked food and before handling food that is ready to eat,
  - after smoking, and
  - after using a tissue or touching their hair, face or body.
- Do not smoke where food is stored, prepared or served (workplaces usually have a designated smoking area outside the building).
- Avoid wearing jewelry or watches.

Q. What is an allergen?

ANS. An allergen is a substance that can cause illness through eating it or having contact with it.

Q. What are the benefits of an HACCP-based food safety program?

ANS. HACCP is the most effective way to ensure food safety.

1. It offers a simple, systematic approach to identifying and controlling hazards at all stages of the process, from purchase of ingredients through to sales or service.
2. Rather than relying on end-product testing, HACCP prevents a food safety problem from occurring in the first place.
3. HACCP provides a business with confidence in its food products and customers can feel secure about food safety standards.
4. By identifying and controlling hazards that can affect food safety and by improving food-processing systems, the overall quality of food products is enhanced.
5. HACCP helps the business comply with Australian and international legal requirements.

6. The implementation of HACCP can support a "due diligence" defense for the business if food safety problems do occur (that is, a legal defense that shows a business has taken all reasonable precautions and exercised all due diligence to avoid the occurrence of a food-borne incident).

Q. Which areas of typical CCPs require steps to prevent, eliminate or reduce hazards in a food service environment?

ANS. Receiving, mixing, preparing, cooking, cooling, re-heating, hot and cold holding are areas of typical CCPs requiring steps to prevent, eliminate or reduce hazards.

Q. What are the basic steps in cleaning dishes, utensils and equipment?

ANS.

1. Put away food before starting to clean floors and walls.
2. Rinse by pre-soaking, sweeping and wiping down a surface.
3. Wash utensils in clean hot water (at about 60 °C), using a suitable detergent and brushes.
4. Rinse in very hot water (at least 82 °C) or use a chemical sanitizer.
5. Air-dry.
6. Dismantle equipment and wash the parts in a sink and wipe down fixed parts with a clean cloth.

Q. Who should be part of a HACCP team?

ANS. An HACCP team may consist of executives, managers and heads.

Q. What is the purpose of the HACCP team?

ANS. It has to develop HACCP plans (FSP) and implement them into the operation.

Q. How is the food safety program communicated?

ANS. Regular communication with staff; on- and off-the-job training sessions; presenting the information in a way that is easy and simple to follow.

Q. What is the difference between a control measure, a CCP and a critical limit?

ANS. A control measure is an action or procedure that will reduce, prevent or eliminate a potential hazard. A CCP is a step at which a control measure is applied. A control limit is a maximum and/or minimum value for controlling a chemical, biological or physical parameter.

Q. What is the greatest hazard in the food industry today?

ANS. Microbiological contamination is the chief hazard in the food industry. However, chemical and physical hazards should not be overlooked.

Q. Does end-product testing of bacteria play a role in HACCP?

ANS. Testing for microbes can play a valuable role in confirming the HACCP system is working properly. Testing is also useful in profiling and tracking products and processes. However, testing for microbes at the end of a process is not effective in identifying and eliminating contamination.

Q. Write examples of three types of contaminations (hazards)?

ANS.

1. Biological—bacteria, virus, parasites, fungi and natural toxins.
2. Chemical—cleaners, sanitizers, toxic metal from non-food service grade utensils and cookware and pesticides.
3. Physical—foreign objects like hair, glass, paper and metal shavings.

Q. In which four ways does food become contaminated?

ANS.

1. Time-temperature control: TCS foods are left in the danger zone for more than 4 hours.
2. Cross contamination: contaminants cross to a food that is not going to be cooked any further.
3. Poor personal hygiene: food handlers cause the food-borne illness.
4. Poor cleaning and sanitization.

Q. Why HACCP?

ANS. India is a signatory to WTO. The WTO Agreement on Sanitary and Phytosanitary (SPS) Measures makes it obligatory to adopt the standards, guidelines and recommendations issued by the Codex Alimentarius Commission which advocates the adoption of HACCP. The Indian Standard on "Food Hygiene—Hazard Analysis and Critical Control Point (HACCP)— System and Guidelines for Its Application" IS 15000:1998 is technically equivalent to the Codex document on the subject. For the food industry in India, adoption of HACCP is becoming imperative to reach global standards, demonstrate compliance to regulations/customer requirements besides providing safer food to the country's millions.

Q. Is there any standard published by ISO on HACCP?

ANS. At present, there is no ISO standard available for HACCP certification.

Q. What are the details about HACCP?

ANS. HACCP involves a system approach to identification of hazards, assessment of chances of occurrence of hazards during each phase, raw material procurement, manufacturing, distribution, usage of food products and defining the measures for hazard control. In doing so, the many drawbacks prevalent in the inspection approach are avoided and HACCP overcomes shortcomings of reliance only on microbial testing.

Q. Who can implement HACCP?

ANS. HACCP enables producers, processors, distributors, exporters, etc. of food products to utilize technical resources efficiently and in a cost-effective manner to ensure food safety. Food inspection too would be more systematic and, therefore, hassle-free. It would involve deployment of some additional finances initially but this would be more than compensated in the long run through consistently better quality and hence better prices and returns.

Q. What is HACCP certification?

ANS. BIS offers two certification schemes to the food industry:

1. Food Safety Certification (HACCP) against IS 15000:1998.
2. HACCP-based Quality Management Systems Certification provides for twin certification through one audit, that is, Certification of QMS against IS/ISO 9001:2008 and Certification of HACCP against IS 15000:1998.

Q. Which temperature range represents the temperature danger zone?

ANS. A temperature range of 41 °F to 135 °F represents the temperature danger zone.

Q. How long should employees wash their hands?

ANS. Employees should wash their hands for twenty seconds.

Q. Which of the following provides important information about the safety of using a chemical?

ANS. The Material Safety Data Sheet (MSDS) provides such information.

Q. What are the recommended holding temperatures for cold and hot foods based on the *Food Code*?

ANS. Based on the *Food Code*, the recommended holding temperatures for cold foods is 41 °F or below and for hot foods it is 135 °F or above.

Q. Can the HACCP decision tree be represented in a diagram?

ANS. Yes.

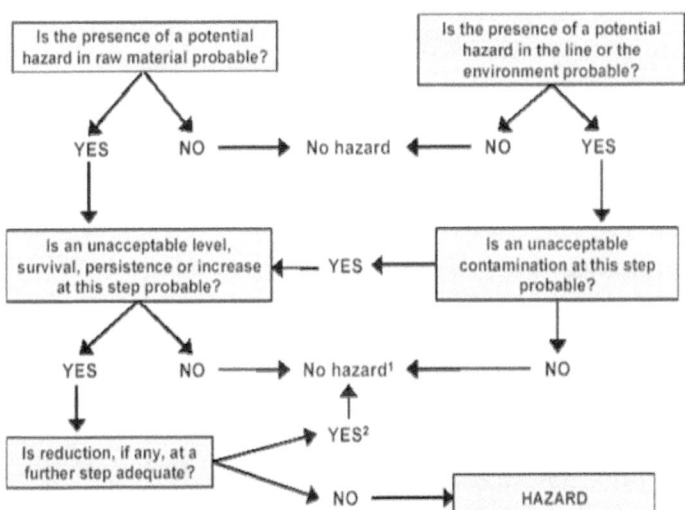

Fig. 33: The HACCP decision tree.

1. Not a hazard to be controlled at this step
2. Thus, reduction step becomes CCP

Q. Can the CCP decision tree be represented in a diagram?
ANS. Yes.

Fig. 34: The CCP decision tree.

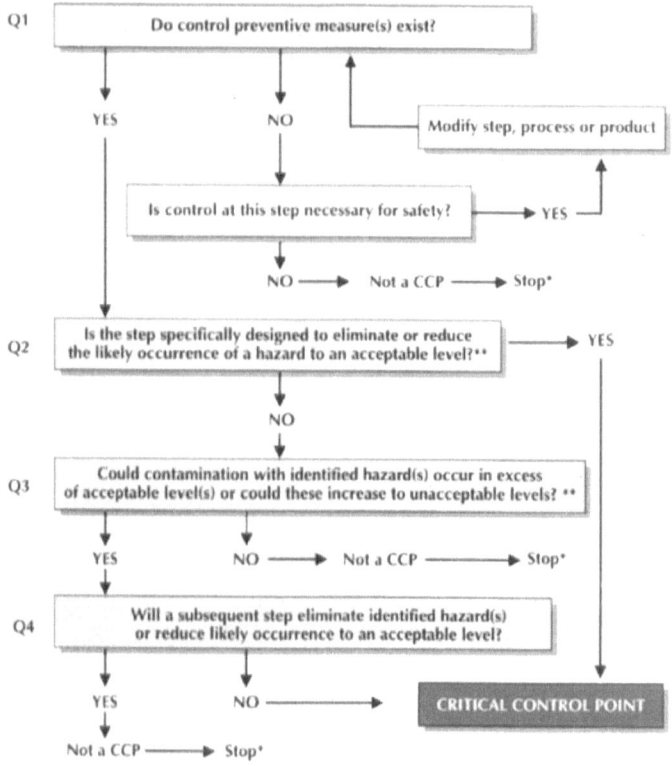

*Proceed to next identified hazard in the described process
** Acceptable and unacceptable levels need to be defined within the overall objectives in identifying the CCPs of HACCP plan.

Q. What is a hazard analysis register?
ANS.

Table 33: An example of a hazard analysis register.

Step No.	Process Step	Hazard Type	Potential Hazard and Cause	Likelihood	Severity	Risk / Hazard	Significant Hazard	Control Measure
1.								

Q. How can control measures be selected and assessed?

ANS. In selecting control measures the following criteria can be considered:

Table 34: Criteria for selecting control measures.

Sr. No	Criteria
	Its effect on identified food safety hazards relative to the strictness applied.
	Its feasibility for monitoring (for example, ability to be monitored in a timely manner to enable immediate corrections).
	Its place within the system relative to other control measures.
	The likelihood of failure in the functioning of a control measure or significant processing variability.
	The severity of the consequence(s) in the case of failure in its functioning.
	Whether the control measure is specifically established and applied to eliminate or significantly reduce the level of hazard(s).
	Synergistic effects (that is, interaction that occurs between two or more measures resulting in their combined effect being higher than the sum of their individual effects).

Table 35

Assessment of Control Measures										Total	Management	
Step No.	Process Step	Product Category	Significant Hazard	Control Measures	Assessment Criteria (Levels of effectiveness)						HACCP Plan	OPRP
					1:Low/2:Medium/3:High							
					A	B	C	D	E	F	G		More than 13	Less than 13

Q. What is the format for a HACCP plan?

ANS.

Table 36: The format for a HACCP plan.

CCP/ OPRP No.	
Process Step No.	
Hazards	
Control Measures	
Critical Limits	
Monitoring Procedures	
What	
How	
When	
Where	
Who	
Corrective Action	
Record Name	
Verification	
What	
How	
When	
Who	

or

1	2	3	4	5	6	7	8	9	10
OPRP/ CCP No.	Significant Hazards	Control Measure	Monitoring				Corrective Action (s)	Records	Verification
			What	How	When	Who			

SAMPLE PAPER FOR OFFICER (CANE)

Q. What is the botanical name of sugarcane?

ANS. *Saccharum officinerum*

Q. What are ten early group and general group varieties?

ANS. Early group:

1. Co 0238
2. Co 0239
3. Co 0118
4. Co 98014
5. Co 05009
6. CoSe 94184
7. CoS 8436
8. CoS 96268
9. CoS 8272
10. CoSe 98231

General group:

Co 5011
Co 0124
CoS 7250
CoS 8276
CoS 767
CoS 97264
CoS 8279
U.P. 0097
CoH 167
CoS 98259

Q. What are the five most harmful insect pests?

ANS. Most common and harmful insect pests are as under:

1. Termite (*odontotermes obesus*)
2. White grub (*holotrichia sps.*)
3. Early shoot borer (*chilo infuscatelus*)
4. Top borer (*tryporyza nivella*)
5. Root borer (*emmalocera depressella*)

Q. What are the five most common diseases in sugarcane?

ANS. The most common diseases in sugarcane are as under:

1. Red rot (*colletotrichum falcatum*)
2. Wilt (*fusarium moniliforme*)

3. Smut (*ustilago centenary*)
4. Grassy shoot (phytoplasmic disease)
5. Top rot (*fusarium moniliforme*)

Q. What is the balanced dose of nutrients (NPK) kg/hectare.
ANS. The balanced dose of NPK is as under:
1. Nitrogen: 150–180 kg/hectare
2. Phosphorus: 80 kg/hectare
3. Potassium: 60 kg/hectare

Q. What are the fungicides used for seed treatment?
ANS. Rexil or Bavistin are both mercury-based fungicides used for seed treatment.

Q. What is the quantity of cane seed/acre?
ANS. The seed cane quantity may fluctuate with the thickness of cane. For instance, thin cane seed requirement can be fulfilled by 25–30 quintals/acre, whereas thick cane seed requirement is fulfilled with 35–45 quintals/acre.

Q. How many meters make an acre and a hectare?
ANS. One acre has 4,000 meters and one hectare has 10,000 meters.

Q. What are the benefits of foliar application of nutrients?
ANS. Foliar application of nutrients fulfill the requirements of plants instantly and 90–95% fertilizer is absorbed by plant leaves, whereas in topdressing, fertilizer utilization is a maximum 50%. Secondly, the application of nutrients in combination with insecticide also helps control sap-sucker or leaf-cutter insect pests.

Q. What are the parasites of top borer, pyrilla and woolly aphid?
ANS. The parasite names are as under:
1. Top borer—*trichogramma chilonis*
2. Pyrilla—*epiricania melanoleuca*
3. Woolly aphid—*dipha*

Q. What is IPM?
ANS. Integrated Pest Management (IPM) is an ecosystem-based strategy that focuses on long-term prevention of pests or their damage through a combination of techniques such as biological control, habitat manipulation, modification of cultural practices and use of resistant varieties. Pesticides are used only after monitoring indicates they are needed according to established guidelines, and treatments are made with the goal of removing only the target organism. Pest control materials are selected and applied in a manner that minimizes risks to human health, beneficial and non-targeted organisms, and the environment.

Q. What is photosynthesis in plants?
ANS. Photosynthesis functions as a counterbalance to respiration; it takes in the carbon dioxide produced by all breathing organisms and reintroduces oxygen into the atmosphere.

Photosynthesis is written as under:

$$6\ CO_2 + 12\ H_2O + \text{Light Energy} \rightarrow C_6H_{12}O_6 + 6\ O_2 + 6\ H_2O$$

Here, six molecules of carbon dioxide (CO_2) combine with 12 molecules of water (H_2O) using light energy. The end result is the formation of a single carbohydrate molecule ($C_6H_{12}O_6$, or glucose) along with six molecules each of breathable oxygen and water.

Q. What are some green manure crops?
ANS. Green gram, black gram, sun hemp and soya bean.

Q. Why must the upper one-third portion of cane be used in planting?
ANS. When this portion is used in planting, it results in much better and faster germination with less disease infestation.

Q. What is the benefit of trash mulching in ratoon crops?
ANS. The following benefits are derived from trash mulching:
1. Control of weeds;
2. Reduction in the requirement of water;
3. Fulfillment of the organic carbon content in soil; and
4. Control of soil health degradation.

Q. What is SSI?

ANS. It refers to the practice of raising young cane plants in a nursery using small chips taken from the cane with the remaining cane being used in factory supply. The seed requirement in this method is very less. After cutting the chips of the bud, prepare a nursery in a tray containing coco pith and vermi-compost, transplanting the seedlings while still young (twenty-five to thirty-five days) with wide spacing between rows and plants (120 × 60 cm). This practice yields higher number of millable cane with more weight and greater suitability for rapid multiplication of new improved cane variety.

Q. What is RBS planting technique?

ANS. In this technique, seed cane is cut in single bud and from these buds it is established on bed. These seedlings transplanted in the fields is the RBS technique.

Q. What is soil fertility?

ANS. The ability of the soil to provide all essential plant nutrients in available form is called soil fertility.

Q. What is soil productivity?

ANS. Soil productivity is the capacity of soil to produce plants under a specified program of management.

Q. What are nutrients?

ANS. Nutrients are essential for the growth of any plant. They are divided into three groups:

1) major, primary or macro nutrients, 2) secondary nutrients and 3) micro or trace nutrients.

Q. What are the groups of nutrients and how many nutrients are included in one group?

ANS. Primary group: This group includes **N, P** and **K** which are required by plants in a huge quantity.

Secondary group: This group has three nutrients, namely, **Ca, Mg** and **S**. They are required right from the beginning of the plant growth but in lesser quantity than primary nutrients.

Micro/Trace nutrients: This group includes **Zn, Fe, Cu, Mn, Mo, Bo, Cl and silica**. These elements are required in a very limited quantity by plants and they act as catalyst or accelerate the enzyme activities in metabolic processes of plants.

Q. How can termites be controlled without chemical application?

ANS. We can control termites by digging a pit 1 × 1 feet deep and wide along the border and in the corners of fields and filled by raw (fresh) cow dung for two-three weeks. After two-three weeks, all termites gather in the cow dung within the pit. This cow dung should be extracted from the pit and doused in pond/water or burnt with kerosene. This activity can be repeated three to four times in a field within the crop cycle.

Q. What is the name of the fungus that controls termites and white grub?

ANS. *Buevaria bassiana* and *metaryzium anasopli* control both insect pests without using chemicals, but its culture must be repeated at least two times in a crop cycle.

Q. What is the parentage of Co 0238, CoJ 85, CoS 8436, CoJ 88, Co 5011 and CoS 7250 with the release year?

ANS. A) Early group varieties:

Co 0238 = CoLk 8102 × Co 775 (year of release: 2009)
CoJ 85 = Q 63 × CoJ 70 (year of release: 2000)
CoS 8436 = MS 68/47 × Co 1148 (year of release: 1986)

B) General group varieties:

Co 5011 = CoS 8436 × Co 89003 (year of release: 2014)
CoJ 88 = CoJ 82315 × Co 1148 (year of release: 2002)
CoS 7250 = CoS 8436 × Co 775 (year of release: 2009)

Q. What are weeds?

ANS. Weeds are any plants or vegetation interfering with the objective of the requirements of people. They are plants growing where they are not desired. A plant whose economic value has not been discovered and is noxious, useless, unwanted or poisonous is a weed.

Q. Why are intercropping and crop rotation important for sugarcane?

ANS. Due to intercropping, sugarcane gets more inter-culturing practice and soil has pulverization for good development of root system. Some crops help control insect pest infestation or secondary infestation of disease. Crop rotation is also beneficial for increasing yield and preventing degradation in variety.

Q. What are the bonding limits of all types of farmers?

ANS. The details of bonding limits are as under:

1. Marginal farmers—1 hectare (maximum 750 quintals or/and in case of yield enhancement maximum of 1,000 quintals)

2. Small farmers—2 hectares (maximum 1,500 quintals or/and in case of yield enhancement maximum of 2,000 quintals)

3. General farmer—5 hectares (maximum 3,750 quintals or/and in case of yield enhancement maximum 5,000 quintals)

Maximum quality is based on cane area (hectare) × 750 quintals and in case of yield enhancement, maximum 5,000 quintals or whichever is less shall be taken.

Q. What are the facilities available to marginal growers in cane-bonding policy?

ANS. Priority shall be given to marginal sugarcane farmers for supplying their plant and ratoon cane within forty-five days of the commencement of the mill and from February 1 within forty-five days. Farmers having sugarcane bond of four bullock carts (60 quintals) shall be deemed marginal farmers.

Q. How is the basic quota of a farmer to be calculated?

ANS. Basic quota calculation should depend on supply over two regular years. For new growers, it is equal to the concerned out-center or the mill gate average supply of the previous crushing season.

Basic quota = Two years supply/2.

Q. What is brix in sugarcane?

ANS. A reading that can be measured, it refers to the available solids in cane juice. It is always indicated in percentage.

Q. What is pol in cane?

ANS. A reading that can be measured, it refers to the value of sucrose in sugarcane juice. It can be measured by Polari meter.

REFERENCES

1. Anderson and Bowen (1990). *Sugarcane Nutrient Management in Leaf Analysis.*
2. Dr. B S Tomer (2005). *Ganna Prashanottary.*
3. H.Rostron, (1972). SASTA.
4. ICAR. Nutrients Leaflet.
5. IISR, Lucknow. Agricultural Implements Leaflets.
6. ISO 19001-Guideline for Auditing Management System.
7. ISO 22000–2005.
8. ISO 31000 Risk Management.
9. ISO 9000-Quality Management System Fundamental & Vocabulary.
10. ISO Standard 9001:2015.
11. KALRO-SRI (2015). Monthly newslatter.
12. McCray et al. (2006). SS AGR -128.
13. McCray & Mylaravapu (2010). SS AGR-335.
14. N C Verma (2005). *System of Technical Control for Cane Sugar Factories.*
15. NSI, Kanpur. Sugarcane Quality System.
16. SBIRC, Karnal. Varieties leaflet (01/2010).
17. Solomon, S. 1997
18. The BioScan 9(2): 2014
19. TEIL market research (2014).
20. TNAU- Organic Farming (2012).
21. UPCSR, Shahajanpur. Varieties Leaflet (2014).
22. Uttar Pradesh Bonding Policy (2015) (C.C. U.P. Govt.).

SABZI SEED STORE
DEALS IN QUALITY SEED, PESTICIDES & SPRAYERS
COURT ROAD MUZAFFARNAGAR (U.P.)
PH. 0131-2436503, sabziseedstore@gmail.com

&

BEST SEED COMPANY
139, INDIRA MARKET OLD SABZI MANDI
DELHI

SANJAY DHINGRA	+91 9557775550
DIRECTOR	+91 9837033203
	+91 7838280022

With Best Compliments from,

LSS LEANING SECURITY SERVICE

73-A, N-Block, Kidwai Nagar, KANPUR-11
Ph. 0512-2643284, Mob. +919935023284, email: leaningsecurity@gmail.com

* We are providing services in following areas:
* Cane Consultancy
* Cane Survey
* Cane Development
* Manpower Management
* Statutory Compliance
* Security Services

BEST COMPLIMENTS FROM

ABDUL SATTAR

NEW CANE VARIETY SEED SUPPLIER TO SUGAR FACTORIES OF U.P. & BIHAR
CHAUDHARY FARM BACHHRAU, TEH.- DHANORA, DISTT-AMROHA (U.P.)
MOB. NO. +919627340456

AVAILABLE SEED CANE VARIETIES ARE:

Co 0118, Co 0238, Co 0239, Co 5011, Co 98014

CoJ 88, CoJ 64

CoSe 94184

CoS 7250, CoS 8272